Hands-On System Programming with Go

Build modern and concurrent applications for Unix and Linux systems using Golang

Alex Guerrieri

BIRMINGHAM - MUMBAI

Hands-On System Programming with Go

Copyright © 2019 Packt Publishing

Commissioning Editor: Richa Tripathi
Acquisition Editor: Shriram Shekhar
Content Development Editor: Tiksha Sarang
Senior Editor: Afshaan Khan
Technical Editor: Sabaah Navlekar
Copy Editor: Safis Editing
Language Support Editor: Storm Mann
Project Coordinator: Prajakta Naik
Proofreader: Safis Editing
Indexer: Rekha Nair
Production Designer: Shraddha Falebhai

First published: July 2019

Production reference: 1040719

Published by Packt Publishing Ltd.
Livery Place
35 Livery Street
Birmingham
B3 2PB, UK.

ISBN 978-1-78980-407-2

www.packtpub.com

To my family, who laid the foundation of the path I am on today

– Alex Guerrieri

Packt>

Packt.com

Subscribe to our online digital library for full access to over 7,000 books and videos, as well as industry leading tools to help you plan your personal development and advance your career. For more information, please visit our website.

Why subscribe?

- Spend less time learning and more time coding with practical eBooks and videos from over 4,000 industry professionals

- Improve your learning with Skill Plans built especially for you

- Get a free eBook or video every month

- Fully searchable for easy access to vital information

- Copy and paste, print, and bookmark content

Did you know that Packt offers eBook versions of every book published, with PDF and ePub files available? You can upgrade to the eBook version at www.packt.com and, as a print book customer, you are entitled to a discount on the eBook copy. Get in touch with us at customercare@packtpub.com for more details.

At www.packt.com, you can also read a collection of free technical articles, sign up for a range of free newsletters, and receive exclusive discounts and offers on Packt books and eBooks.

Contributors

About the author

Alex Guerrieri is a software developer who specializes in backend and distributed systems. Go has been his favorite tool for the job since first using it in 2013. He holds a degree in computer science engineering and has been working in the field for more than 6 years. Originally from Italy, where he completed his studies and started his career as a full-stack developer, he now lives in Spain, where he previously worked in two start-ups—source{d} and Cabify. He is now working for three companies—as a software crafter for BBVA, one of the biggest Spanish banks; as a software architect for Security First, London, a company focusing on digital and physical security; and as a cofounder of DauMau, the company behind Vidsey, software that generates videos in a procedural manner.

> *To all the people in my life who supported me and put up with me throughout this endeavor of writing this book, thank you.*

About the reviewers

Corey Scott is a principal software engineer currently living in Melbourne, Australia. He has been programming professionally since 2000, with the last 5 years spent building large-scale distributed services in Go.

A blogger on a variety of software-related topics, he is passionate about designing and building quality software. He believes that software engineering is a craft that should be honed, debated, and continuously improved. He takes a pragmatic, non-zealous approach to coding, and is always up for a good debate about software engineering, continuous delivery, testing, or clean coding.

Janani Selvaraj is currently working as a data analytics consultant for Gaddiel Technologies, Trichy, where she focuses on providing data analytics solutions for start-up companies. Her previous experience includes training and research development in relation to data analytics and machine learning.

She has a PhD in environmental management and has more than 5 years' research experience with regard to statistical modeling. She is also proficient in a number of programming languages, including R, Python, and Go.

She reviewed a book entitled *Go Machine Learning Projects*, and also coauthored a book entitled *Machine Learning Using Go*, published by *Packt Publishing*.

Arun Muralidharan is a software developer with over 9 years' experience as a systems developer. Distributed system design, architecture, event systems, scalability, performance, and programming languages are some of the aspects of a product that interest him the most. Professionally, he spends most of his time coding in C++, Python, and C (and perhaps Go in the near future). Away from his job, he also develops software in Go and Rust.

I would like to take this opportunity to thank my family, who have provided me with unconditional support over the years.

Packt is searching for authors like you

If you're interested in becoming an author for Packt, please visit `authors.packtpub.com` and apply today. We have worked with thousands of developers and tech professionals, just like you, to help them share their insight with the global tech community. You can make a general application, apply for a specific hot topic that we are recruiting an author for, or submit your own idea.

Table of Contents

Preface

This book will provide good, in-depth explanations of various interesting Go concepts. It begins with Unix and system programming, which will help you understand what components the Unix operating system has to offer, from the kernel API to the filesystem, and allow you to familiarize yourself with the basic concepts of system programming.

Next, it moves on to cover the application of I/O operations, focusing on the filesystem, files, and streams in the Unix operating system. It covers many topics, including reading from and writing to files, among other I/O operations.

This book also shows how various processes communicate with one another. It explains how to use Unix pipe-based communication in Go, how to handle signals inside an application, and how to use a network to communicate effectively. Also, it shows how to encode data to improve communication speed.

The book will, toward the end, help you to understand the most modern feature of Go—concurrency. It will introduce you to the tools the language has, along with sync and channels, and how and when to use each one.

Who this book is for

This book is for developers who want to learn system programming with Go. Although no prior knowledge of Unix and Linux system programming is necessary, some intermediate knowledge of Go will help you to understand the concepts covered in the book.

What this book covers

Chapter 1, *An Introduction to System Programming*, introduces you to Go and system programming and provides some basic concepts and an overview of Unix and its resources, including the kernel API. It also defines many concepts that are used throughout the rest of the book.

Chapter 2, *Unix OS Components*, focuses on the Unix operating system and the components that you will interact with—files and the filesystem, processes, users and permissions, threads, and others. It also explains the various memory management techniques of the operating system, and how Unix handles resident and virtual memory.

Chapter 3, *An Overview of Go*, takes a look at Go, starting with some history of the language and then explaining, one by one, all its basic concepts, starting with namespaces and the type system, variables, and flow control, and finishing with built-in functions and the concurrency model, while also offering an explanation of how Go interacts and manages its memory.

Chapter 4, *Working with the Filesystem*, helps you to understand how the Unix filesystem works and how to master the Go standard library to handle file path operations, file reading, and file writing.

Chapter 5, *Handling Streams*, helps you to learn about the interfaces for the input and output streams that Go uses to abstract data flows. It explains how they work and how to combine them and best use them without leaking information.

Chapter 6, *Building Pseudo-Terminals*, helps you understand how a pseudo-terminal application works and how to create one. The result will be an interactive application that uses standard streams just as the command line does.

Chapter 7, *Handling Processes and Daemons*, provides an explanation of what processes are and how to handle them in Go, how to start child processes from a Go application, and how to create a command-line application that will stay in the background (a daemon) and interact with it.

Chapter 8, *Exit Codes, Signals, and Pipes*, discusses Unix inter-process communication. It explains how to use exit codes effectively. It shows you how signals are handled by default inside an application, and how to manage them with some patterns for effective signal handling. Furthermore, it explains how to connect the output and input of different processes using pipes.

Chapter 9, *Network Programming*, explains how to use a network to make processes communicate. It explains how network communication protocols work. It initially focuses on low-level socket communication, such as TCP and UDP, before moving on to web server development using the well-known HTTP protocol. Finally, it shows how to use the Go template engine.

Chapter 10, *Data Encoding Using Go*, explains how to leverage the Go standard library to encode complex data structures in order to facilitate process communications. It analyzes both text-based protocols, such as XML and JSON, and binary-based protocols, such as GOB.

Chapter 11, *Dealing with Channels and Goroutines*, explains the basics of concurrency and channels and some general rules that prevent the creation of deadlocks and resource-leaking inside an application.

Chapter 12, *Synchronization with sync and atomic*, discusses the synchronization packages of the `sync` and `sync/atomic` standard libraries, and how they can be used instead of channels to achieve concurrency easily. It also focuses on avoiding the leaking of resources and on recycling resources.

Chapter 13, *Coordination Using Context*, discusses `Context`, a relatively new package introduced in Go that offers a simple way of handling asynchronous operations effectively.

Chapter 14, *Implementing Concurrency Patterns*, uses the tools from the previous three chapters and demonstrates how to use and combine them to communicate effectively. It focuses on the most common patterns used in Go for concurrency.

Chapter 15, *Using Reflection*, explains what reflection is and whether you should use it. It shows where it's used in the standard library and guides you in creating a practical example. It also shows how to avoid using reflection where there is no need to.

Chapter 16, *Using CGO*, explains how CGO works and why and when you should use it. It explains how to use C code inside a Go application, and vice versa.

To get the most out of this book

Some basic knowledge of Go is required to try the examples and to build modern applications.

Each chapter includes a set of questions that will help you to gauge your understanding of the chapter. The answers to these questions are provided in the *Assessments* section of the book. These questions will prove very beneficial for you, as they will help you revisit each chapter at a glance.

Apart from this, each chapter provides you with instructions on how to run the code files, while the GitHub repository of the book provides the requisite details.

Download the example code files

You can download the example code files for this book from your account at www.packt.com. If you purchased this book elsewhere, you can visit www.packt.com/support and register to have the files emailed directly to you.

You can download the code files by following these steps:

1. Log in or register at `www.packt.com`.
2. Select the **SUPPORT** tab.
3. Click on **Code Downloads & Errata**.
4. Enter the name of the book in the **Search** box and follow the onscreen instructions.

Once the file is downloaded, please make sure that you unzip or extract the folder using the latest version of:

- WinRAR/7-Zip for Windows
- Zipeg/iZip/UnRarX for Mac
- 7-Zip/PeaZip for Linux

The code bundle for the book is also hosted on GitHub at `https://github.com/PacktPublishing/Hands-On-System-Programming-with-Go`. In case there's an update to the code, it will be updated on the existing GitHub repository.

We also have other code bundles from our rich catalog of books and videos available at `https://github.com/PacktPublishing/`. Check them out!

Download the color images

We also provide a PDF file that has color images of the screenshots/diagrams used in this book. You can download it here: `https://static.packt-cdn.com/downloads/9781789804072_ColorImages.pdf`.

Code in Action

Visit the following link to check out videos of the code being run: `http://bit.ly/2ZWgJb5`.

Playground examples

In the course of the book you will find many snippets of code followed by a link to `https://play.golang.org`, a service that allows you to run Go applications with some limitations. You can read more about it at `https://blog.golang.org/playground`.

In order to see the full source code of such examples, you need to visit the Playground link. Once on the website, you can press the **Run** button to execute the application. The bottom part of the page will show the output. The following is an example of the code running in the Go Playground:

```
The Go Playground    Run  Format  ■ Imports  Share                    About

 1 package main
 2
 3 import (
 4        "encoding/csv"
 5        "fmt"
 6        "os"
 7        "strconv"
 8 )
 9
10 func main() {
11        const million = 1000000
12        type Country struct {
13                Code, Name string
14                Population int
15        }
16        records := []Country{
17                {Code: "IT", Name: "Italy", Population: 60 * million},
18                {Code: "ES", Name: "Spain", Population: 46 * million},
19                {Code: "JP", Name: "Japan", Population: 126 * million},
20                {Code: "US", Name: "United States of America", Population: 327 * million},
21        }
22        w := csv.NewWriter(os.Stdout)
23        defer w.Flush()
24        for _, r := range records {

IT,Italy,60000000
ES,Spain,46000000
JP,Japan,126000000
US,United States of America,327000000

Program exited.
```

If you want, you have the possibility of experimenting by adding and editing more code to the examples, and then running them.

Conventions used

There are a number of text conventions used throughout this book.

`CodeInText`: Indicates code words in text, database table names, folder names, filenames, file extensions, pathnames, dummy URLs, user input, and Twitter handles. Here is an example: "This type of service includes `load`, which adds a program to memory and prepares for its execution before passing control to the program itself, or `execute`, which runs an executable file in the context of a pre-existing process."

A block of code is set as follows:

```
<meta name="go-import" content="package-name vcs repository-url">
```

Any command-line input or output is written as follows:

```
Enable-WindowsOptionalFeature -Online -FeatureName Microsoft-Windows-
Subsystem-Linux
```

Bold: Indicates a new term, an important word, or words that you see on screen. For example, words in menus or dialog boxes appear in the text like this. Here is an example: "In the meantime, systems started to get distributed, and applications started to get shipped in containers, orchestrated by other system software, such as **Kubernetes**."

 Warnings or important notes appear like this.

 Tips and tricks appear like this.

Get in touch

Feedback from our readers is always welcome.

General feedback: If you have questions about any aspect of this book, mention the book title in the subject of your message and email us at customercare@packtpub.com.

Errata: Although we have taken every care to ensure the accuracy of our content, mistakes do happen. If you have found a mistake in this book, we would be grateful if you would report this to us. Please visit www.packt.com/submit-errata, selecting your book, clicking on the Errata Submission Form link, and entering the details.

Piracy: If you come across any illegal copies of our works in any form on the internet, we would be grateful if you would provide us with the location address or website name. Please contact us at copyright@packt.com with a link to the material.

If you are interested in becoming an author: If there is a topic that you have expertise in, and you are interested in either writing or contributing to a book, please visit authors.packtpub.com.

Reviews

Please leave a review. Once you have read and used this book, why not leave a review on the site that you purchased it from? Potential readers can then see and use your unbiased opinion to make purchase decisions, we at Packt can understand what you think about our products, and our authors can see your feedback on their book. Thank you!

For more information about Packt, please visit `packt.com`.

Section 1: An Introduction to System Programming and Go

This section is an introduction to Unix and system programming. It will help you understand what components Unix operating systems have to offer, from the kernel API to the filesystem, and you will become familiar with system programming's basic concepts.

This section consists of the following chapters:

- Chapter 1, *An Introduction to System Programming*
- Chapter 2, *Unix OS Components*
- Chapter 3, *An Overview of Go*

1
An Introduction to System Programming

This chapter is an introduction to system programming, exploring a range of topics from its original definition to how it has shifted in time with system evolution. This chapter provides some basic concepts and an overview of Unix and its resources, including the kernel and the **application programming interfaces** (**API**). Many of these concepts are defined here and are used in the rest of the book.

The following topics will be covered in this chapter:

- What is system programming?
- Application programming interfaces
- Understanding how the protection ring works
- An overview of system calls
- The POSIX standard

Technical requirements

This chapter does not require you to install any special software if you're on Linux.

If you are a Windows user, you can install the **Windows Subsystem for Linux** (**WSL**). Follow these steps in order to install WSL:

1. Open PowerShell as administrator and run the following:

```
Enable-WindowsOptionalFeature -Online -FeatureName Microsoft-
Windows-Subsystem-Linux
```

2. Restart your computer when prompted.
3. Install your favorite Linux distribution from the Microsoft Store.

Beginning with system programming

Over the years, the IT landscape has shifted dramatically. Multicore CPUs that challenge the Von Neumann machine, the internet, and distributed systems are just some of the changes that occurred in the last 30 years. So, where does system programming stand in this landscape?

Software for software

Let's start with the standard textbook definition first.

System programming (or systems programming) is the activity of programming computer system software. The primary distinguishing characteristic of system programming when compared to application programming is that application programming aims to produce software that provides services directly to the user (for example, a word processor), whereas system programming aims to produce software and software platforms that provide services to other software, and are designed to work in performance-constrained environments, for example, operating systems, computational science applications, game engines and AAA video games, industrial automation, and software as a service applications.

The definition highlights two main concepts of what system applications are as follows:

- Software that is used by other software, not directly by the final user.
- The software is hardware aware (it knows how the hardware works), and is oriented toward performance.

This makes it possible to easily recognize as system software operating system kernels, hardware drivers, compilers, and debuggers, and not as system software, a chat client, or a word processor.

Historically, system programs were created using Assembly and C. Then came the shells and the scripting languages that were used to tie together the functionality offered by system programs. Another characteristic of system languages was the control of the memory allocation.

Languages and system evolution

In the last decade, scripting languages gained popularity to the point at which some had significant performance improvement and entire systems were built with them. For example, let's just think about the V8 Engine for JavaScript and the PyPy implementation of Python, which dramatically shifted the performance of these languages.

Other languages, such as Go, proved that garbage collection and performance are not mutually exclusive. In particular, Go managed to replace its own memory allocator written in C with a native version written in Go in release 1.5, improving it to the point where the performance was comparable.

In the meantime, systems started to get distributed and the applications started to get shipped in containers, orchestrated by other system software, such as **Kubernetes**. These systems are meant to sustain huge throughput and achieve it in two main ways:

- By scaling—augmenting the number or the resources of the machines that are hosting the system
- By optimizing the software in order to be more resource effective

System programming and software engineering

Some of the practices of system programming—such as having an application that is tied to the hardware, performance oriented, and working on an environment that is resource constrained—are an approach that can also be valid when building distributed systems, where constraining resource usage allows the reduction of the number of instances needed. It looks like system programming is a good way of addressing generic software engineering problems.

This means that learning the concept of system programming with regards to using the resource of the machine efficiently—from memory usage to filesystem access—will be useful when building any type of application.

Application programming interfaces

APIs are series subroutine definitions, communication protocols, and tools for building software. The most important aspects of an API are the functionalities that it offers, combined with its documentation, which facilitates the user in the usage and implementation of the software itself in another software. An API can be the interface that allows an application software to use a system software.

An API usually has a specific release policy that is meant to be used by a specific group of recipients. This can be the following:

- Private and for internal use only
- Partner and usable by determined groups only—this may include companies that want to integrate the service with theirs
- Public and available for every user

Types of APIs

We'll see that there are several types of APIs, from the ones used to make different application software work together, to the inner ones exposed by the operating system to other software.

Operating systems

An API can specify how to interface an application and the operating system. For instance, Windows, Linux, and macOS have an interface that makes it possible to operate with the filesystem and files.

Libraries and frameworks

The API related to a software library describes and prescribes (provides instructions on how to use it) how each of its elements should behave, including the most common error scenarios. The behavior and interfaces of the API is usually referred to as **library specification**, while the library is the implementation of the rules described in such specification. Libraries and frameworks are usually language bound, but there are some tools that make it possible to use a library in a different language. You can use C code in Go using CGO, and in Python you can use CPython.

Remote APIs

These make it possible to manipulate remote resources using specific standards for communication that allow different technologies to work together, regardless of the language or platform. A good example is the **Java Database Connectivity (JDBC)** API, which allows querying many different types of databases with the same set of functions, or the Java remote method invocation API (Java RMI), which allows the use of remote functions as if they were local.

Web APIs

Web APIs are interfaces that define a series of specifications about the protocol used, message encoding, and available endpoints with their expected input and output values. There are two main paradigms for this kind of API—REST and SOAP:

- REST APIs have the following characteristics:
    - They treat data as a resource.
    - Each resource is identified by a URL.
    - The type of operation is specified by the HTTP method.
- SOAP protocols have the following characteristics:
    - They are defined by the W3C standard.
    - XML is the only encoding used for messages.
    - They use a series of XML schema to verify the data.

Understanding the protection ring

The **protection ring**, also referred to as **hierarchical protection domains**, is the mechanism used to protect a system against failure. Its name is derived from the hierarchical structure of its levels of permission, represented by concentric rings, with privilege decreasing when moving to the outside rings. Between each ring there are special gates that allow the outer ring to access the inner ring resources in a restricted manner.

Architectural differences

The number and order of rings depend on the CPU architecture. They are usually numbered with decreasing privilege, making ring 0 the most privileged one. This is true for i386 and x64 architecture that use four rings (from ring 0 to ring 3) but it's not true for ARM, which uses reverse order (from EL3 to EL0). Most operating systems are not using all four levels; they end up using a two level hierarchy—user/application (ring 3) and kernel (ring 0).

Kernel space and user space

A software that runs under an operating system will be executed at user (ring 3) level. In order to access the machine resources, it will have to interact with the operating system kernel (that runs at ring 0). Here's a list of some of the operations a ring 3 application cannot do:

- Modify the current segment descriptor, which determines the current ring
- Modify the page tables, preventing one process from seeing the memory of other processes
- Use the LGDT and LIDT instructions, preventing them from registering interrupt handlers
- Use I/O instructions such as in and out that would ignore file permissions and read directly from disk

The access to the content of the disk, for instance, will be mediated by the kernel that will verify that the application has permission to access the data. This kind of negotiation improves security and avoids failures, but comes with an important overhead that impacts the application performance.

Some applications can be designed to run directly on the hardware without the framework provided by an operating system. This is true for real-time systems, where there is no compromise on response times and performance.

Diving into system calls

System calls are the way operating systems provide access to the resources for the applications. It is an API implemented by the kernel for accessing the hardware safely.

Services provided

There are some categories that we can use to split the numerous functions offered by the operating system. These include the control of the running applications and their flow, the filesystem access, and the network.

Process control

This type of services includes `load`, which adds a program to memory and prepares for its execution before passing control to the program itself, or `execute`, which runs an executable file in the context of a pre-existing process. Other operations that belong to this category are as follows:

- `end` and `abort`—the first requires the application to exit while the second forces it.
- `CreateProcess`, also known as `fork` on Unix systems or `NtCreateProcess` in Windows.
- Terminate process.
- Get/set process attributes.
- Wait for time, wait event, or signal event.
- Allocate and free memory.

File management

The handling of files and filesystems belongs to file management system calls. There are *create* and *delete* files that make it possible to add or remove an entry from the filesystem, and `open` and `close` operations that make it possible to gain control of a file in order to execute read and write operations. It is also possible to read and change file attributes.

Device management

Device management handles all other devices but the filesystem, such as frame buffers or display. It includes all operations from the request of a device, including the communication to and from it (read, write, seek), and its release. It also includes all the operations of changing device attributes and logically attaching and detaching them.

Information maintenance

Reading and writing the system date and time belongs to the information maintenance category. This category also takes care of other system data, such as the environment. Another important set of operations that belongs here is the request and the manipulation of processes, files, and device attributes.

Communication

All the network operations from handling sockets to accepting connections fall into the communication category. This includes the creation, deletion, and naming of connections, and sending and receiving messages.

The difference between operating systems

Windows has a series of different system calls that cover all the kernel operations. Many of these correspond exactly with the Unix equivalent. Here's a list of some of the overlapping system calls:

	Windows	Unix
Process control	CreateProcess() ExitProcess() WaitForSingleObject()	fork() exit() wait()
File manipulation	CreateFile() ReadFile() WriteFile() CloseHandle()	open() read() write() close()
File protection	SetFileSecurity() InitializeSecurityDescriptor() SetSecurityDescriptorGroup()	chmod() umask() chown()
Device management	SetConsoleMode() ReadConsole() WriteConsole()	ioctl() read() write()
Information maintenance	GetCurrentProcessID() SetTimer() Sleep()	getpid() alarm() sleep()
Communication	CreatePipe() CreateFileMapping() MapViewOfFile()	pipe() shmget() mmap()

Understanding the POSIX standard

In order to ensure consistency between operating systems, IEEE formalized some standards for operating systems. These are described in the following sections.

POSIX standards and features

Portable Operating System Interface (POSIX) for Unix represents a series of standards for operating system interfaces. The first version dates back to 1988 and covers a series of topics like filenames, shells, and regular expressions.

There are many features defined by POSIX, and they are organized in four different standards, each one focusing on a different aspect of the Unix compliance. They are all named POSIX followed by a number.

POSIX.1 – core services

POSIX.1 is the 1988 original standard, which was initially named POSIX but was renamed to make it possible to add more standards to the family without giving up the name. It defines the following features:

- Process creation and control
- Signals:
    - Floating point exceptions
    - Segmentation/memory violations
    - Illegal instructions
    - Bus errors
    - Timers
- File and directory operations
- Pipes
- C library (standard C)
- I/O port interface and control
- Process triggers

POSIX.1b and POSIX.1c – real-time and thread extensions

POSIX.1b focuses on real-time applications and on applications that need high performance. It focus on these aspects:

- Priority scheduling
- Real-time signals
- Clocks and timers

- Semaphores
- Message passing
- Shared memory
- Asynchronous and synchronous I/O
- Memory locking interface

POSIX.1c introduces the multithread paradigm and defines the following:

- Thread creation, control, and cleanup
- Thread scheduling
- Thread synchronization
- Signal handling

POSIX.2 – shell and utilities

POSIX.2 specifies standards for both a command-line interpreter and utility programs as cd, echo, or ls.

OS adherence

Not all operating systems are POSIX compliant. Windows was born after the standard and it is not compliant, for instance. From a certification point of view, macOS is more compliant than Linux, because the latter uses another standard built on top of POSIX.

Linux and macOS

Most Linux distributions follow the **Linux Standard Base (LSB)**, which is another standard that includes POSIX and much more, focusing on maintaining the inter-compatibility between different Linux distributions. It is not considered officially compliant because the developers didn't go into the process of certification.

However, macOS became fully compatible in 2007 with the Snow Leopard distribution, and it has been POSIX-certified since then.

Windows

Windows is not POSIX compliant, but there are many attempts to make it so. There are open source initiatives such as Cygwin and MinGW that provide a less POSIX-compliant development environment, and support C applications using the Microsoft Visual C runtime library. Microsoft itself has made some attempts at POSIX compatibility, such as the Microsoft POSIX subsystem. The latest compatibility layer made by Microsoft is the Windows Linux Subsystem, which is an optional feature that can be activated in Windows 10 and has been well received by developers (including myself).

Summary

In this chapter, we saw what system programming means—writing system software that has some strict requirements, such as being tied to the hardware, using a low-level language, and working in a resource-constrained environment. Its practices can be really useful when building distributed systems that normally require optimizing resource usage. We discussed APIs, definitions that allows software to be used by other software, and listed the different types—the ones in the operating system, libraries and frameworks, and remote and web APIs.

We analyzed how, in operating systems, the access to resources is arranged in hierarchical levels called **protection rings** that prevent uncontrolled usage in order to improve security and avoid failures from the applications. The Linux model simplifies this hierarchy to just two levels called *user* and *kernel* space. All the applications are running in the user space, and in order to access the machine's resources they need the kernel to intercede.

Then we saw one specific type of API called **system calls** that allows the applications to request resources to the kernel, and mediates process control, access and management of files, and devices and network communications.

We gave an overview of the POSIX standard, which defines Unix system interoperability. Among the features defined, there are also the C API, CLI utilities, shell language, environment variables, program exit status, regular expressions, directory structures, filenames, and command-line utility API conventions.

In the next chapter, we will explore the Unix operating system resources such as the filesystem and the Unix permission model. We will look at what processes are, how they communicate with each other, and how they handle errors.

Questions

1. What is the difference between application and system programming?
2. What is an API? Why are APIs so important?
3. Could you explain how protection rings work?
4. Can you make some examples of what cannot be done in user space?
5. What's a system call?
6. Which calls are used in Unix to manage a process?
7. Why is POSIX useful?
8. Is Windows POSIX compliant?

Unix OS Components

2

This chapter will be focusing on Unix OS and on the components that the user will interact with: files and filesystems, processes, users and permissions, and so on. It will also explain some basic process communication and how system program error handling works. All these parts of the operating system will be the ones we will be interacting with when creating system applications.

The following topics will be covered in this chapter:

- Memory management
- Files and filesystems
- Processes
- Users, groups, and permissions
- Process communications

Technical requirements

In the same way as the previous chapter, this one does not require any software to be installed: any other POSIX-compliant shell is enough.

You could choose, for instance, Bash (`https://www.gnu.org/software/bash/`), which is recommended, Zsh (`http://www.zsh.org/`), or fish (`https://fishshell.com/`).

Memory management

The operating system handles the primary and secondary memory usage of the applications. It keeps track of how much of the memory is used, by which process, and what parts are free. It also handles allocation of new memory from the processes and memory de-allocation when the processes are complete.

Techniques of management

There are different techniques for handling memory, including the following:

- **Single allocation**: All the memory, besides the part reserved for the OS, is available for the application. This means that there can only be one application in execution at a time, like in **Microsoft Disk Operating System (MS-DOS)**.
- **Partitioned allocation**: This divides the memory into different blocks called partitions. Using one of these blocks per process makes it possible to execute more than one process at once. The partitions can be relocated and compacted in order to obtain more contiguous memory space for the next processes.
- **Paged memory**: The memory is divided into parts called frames, which have a fixed size. A process' memory is divided into parts of the same size called **pages**. There is a mapping between pages and frames that makes the process see its own virtual memory as contiguous. This process is also known as **pagination**.

Virtual memory

Unix uses the paged memory management technique, abstracting its memory for each application into contiguous virtual memory. It also uses a technique called swapping, which extends the virtual memory to the secondary memory (hard drive or **solid state drives (SSD)**) using a swap file.

When memory is scarce, the operating system puts pages from processes that are sleeping in the swap partition in order to make space for active processes that are requesting more memory, executing an operation called **swap-out**. When a page that is in the swap file is needed by a process in execution it gets loaded back into the main memory for executing it. This is called **swap-in**.

The main issue of swapping is the performance drop when interacting with secondary memory, but it is very useful for extending multitasking capabilities and for dealing with applications that are bigger than the physical memory, by loading just the pieces that are actually needed at a given time. Creating memory-efficient applications is a way of increasing performance by avoiding or reducing swapping.

The `top` command shows details about available memory, swap, and memory consumption for each process:

- `RES` is the physical primary memory used by the process.
- `VIRT` is the total memory used by the process, including the swapped memory, so it's equal to or bigger than `RES`.
- `SHR` is the part of `VIRT` that is actually shareable, such as loaded libraries.

Understanding files and filesystems

A filesystem is a method used to structure data in a disk, and a file is the abstraction used for indicating a piece of self-contained information. If the filesystem is hierarchical, it means that files are organized in a tree of directories, which are special files used for arranging stored files.

Operating systems and filesystems

Over the last 50 years, a large number of filesystems have been invented and used, and each one has its own characteristics regarding space management, filenames and directories, metadata, and access restriction. Each modern operating system mainly uses a single type of filesystem.

Linux

Linux's **filesystem (FS)** of choice is the **extended filesystem (EXT)** family, but other ones are also supported, including XFS, **Journaled File System (JFS)**, and **B-tree File System (Btrfs)**. It is also compatible with the older **File Allocation Table (FAT)** family (FAT16 and FAT32) and **New Technology File System (NTFS)**. The filesystem most commonly used remains the latest version of EXT (EXT4), which was released in 2006 and expanded its predecessor's capacities, including support for bigger disks.

macOS

macOS uses the **Apple File System** (**APFS**), which supports Unix permission and has journaling. It is also metadata-rich and case-preserving, while being a case-insensitive filesystem. It offers support for other filesystems, including HFS+ and FAT32, supporting NTFS for read-only operations. To write to such a filesystem, we can use an experimental feature or third-party applications.

Windows

The main filesystem used by Windows is NTFS. As well as being case-insensitive, the signature feature that distinguishes Windows FS from others is the use of a letter followed by a colon to represent a partition in paths, combined with the use of backslash as a folder separator, instead of a forward slash. Drive letters, and the use of C for the primary partition, comes from MS-DOS, where A and B were reserved drive letters used for floppy disk drives.

Windows also natively supports other filesystems, such as FAT, which is a filesystem family that was very popular between the late seventies and the late nineties, and **Extended File Allocation Table** (**exFAT**), which is a format developed by Microsoft on top of FAT for removable devices.

Files and hard and soft links

Most files are regular files, containing a certain amount of data. For instance, a text file contains a sequence of human-readable characters represented by a certain encoding, while a bitmap contains some metadata about the size and the bit used for each pixel, followed by the content of each pixel.

Files are arranged inside directories that make it possible to have different namespaces to reuse filenames. These are referred to with a name, their human-readable identifier, and organized in a tree structure. The path is a unique identifier that represents a directory, and it is made by the names of all the parents of the directory joined by a separator (/ in Unix, \ in Windows), descending from the root to the desired leaf. For instance if a directory named a is located under another named b, which is under one called c, it will have a path that starts from the root and concatenates all the directories, up to the file: /c/b/a.

When more than one file points to the same content, we have a **hard link**, but this is not allowed in all filesystems (for example, NTFS and FAT). A **soft link** is a file that points to another soft link or to a hard link. Hard links can be removed or deleted without breaking the original link, but this is not true for soft links. A **symbolic link** is a regular file with its own data that is the path of another file. It can also link other filesystems or files and directories that do not exist (that will be a broken link).

In Unix, some resources that are not actually files are represented as files, and communication with these resources is achieved by writing to or reading from their corresponding files. For instance, the /dev/sda file represents an entire disk, while /dev/stdout, dev/stdin, and /dev/stderr are standard output, input, and error. The main advantage of *Everything is a file* is that the same tools that can be used for files can also interact with other devices (network and pipes) or entities (processes).

Unix filesystem

The principles contained in this section are specific to the filesystems used by Linux, such as EXT4.

Root and inodes

In Linux and macOS, each file and directory is represented by an **inode**, which is a special data structure that stores all the information about the file except its name and its actual data.

Inode 0 is used for a null value, which means that there is no inode. Inode 1 is used to record any bad block on the disk. The root of the hierarchical structure of the filesystem uses inode 2. It is represented by /.

From the latest Linux kernel source, we can see how the first inodes are reserved. This is shown as follows:

```
#define EXT4_BAD_INO 1 /* Bad blocks inode */
#define EXT4_ROOT_INO 2 /* Root inode */
#define EXT4_USR_QUOTA_INO 3 /* User quota inode */
#define EXT4_GRP_QUOTA_INO 4 /* Group quota inode */
#define EXT4_BOOT_LOADER_INO 5 /* Boot loader inode */
#define EXT4_UNDEL_DIR_INO 6 /* Undelete directory inode */
#define EXT4_RESIZE_INO 7 /* Reserved group descriptors inode */
#define EXT4_JOURNAL_INO 8 /* Journal inode */
```

This link is the source for the preceding code block: `https://elixir.bootlin.com/linux/latest/source/fs/ext4/ext4.h#L212`.

Directory structure

In Unix filesystems, there is a series of other directories under the root, each one used for a specific purpose, making it possible to maintain a certain interoperability between different operating systems and enabling compiled software to run on different OSes, making the binaries portable.

This is a comprehensive list of the directories with their scope:

Directory	Description
/bin	Executable files for all users
/boot	Files for booting the system
/dev	Device drivers
/etc	Configuration files for applications and system
/home	Home directory for users
/kernel	Kernel files
/lib	Shared library files and other kernel-related files
/mnt	Temporary filesystems, from floppy disks and CDs to flash drives
/proc	File with process numbers for active processes
/sbin	Executable files for administrators
/tmp	Temporary files that should be safe to delete
/usr	Administrative commands, shared files, library files, and others
/var	Variable-length files (logs and print files)

Navigation and interaction

While using a shell, one of the directories will be the **working directory**, when paths are relative (for example, `file.sh` or `dir/subdir/file.txt`). The working directory is used as a prefix to obtain an absolute one. This is usually shown in the prompt of the command line, but it can be printed with the `pwd` command (print working directory).

The `cd` (change directory) command can be used to change the current working directory. To create a new directory, there's the `mkdir` (make directory) command.

To show the list of files for a directory, there's the `ls` command, which accepts a series of options, including more information (`-l`), showing hidden files and directories (`-a`), and sorting by time (`-t`) and size (`-S`).

There is a series of other commands that can be used to interact with files: the `touch` command creates a new empty file with the given name, and to edit its content you can use a series of editors, including vi and nano, while `cat`, `more`, and `less` are some of the commands that make it possible to read them.

Mounting and unmounting

The operating system splits the hard drive into logical units called partitions, and each one can be a different file system. When the operating system starts, it makes some partitions available using the `mount` command for each line of the `/etc/fstab` file, which looks more or less like this:

```
# device # mount-point # fstype # options # dumpfreq # passno
/dev/sda1     /           ext4    defaults     0          1
```

This configuration mounts `/dev/sda1` to `/disk` using an `ext4` filesystem and default options, no backing up (0), and root integrity check (1). The `mount` command can be used at any time to expose partitions in the filesystem. Its counterpart, `umount`, is needed to remove these partitions from the main filesystem. The empty directory used for the operation is called **mount point**, and it represents the root under which the filesystem is connected.

Processes

When an application is launched, it becomes a process: a special instance provided by the operating system that includes all the resources that are used by the running application. This program must be in **Executable and Linkable Format (ELF)**, in order to allow the operating system to interpret its instructions.

Process properties

Each process is a five-digit identifier **process ID (PID)**, and it represents the process for all its life cycle. This means that there cannot be two processes with the same PID at the same time. Their uniqueness makes it possible to access a specific process by knowing its PID. Once a process is terminated, its PID can be reused for another process, if necessary.

Similar to PID, there are other properties that characterize a process. These are as follows:

- **PPID**: The parent process ID of the process that started this process
- **Nice number**: Degree of friendliness of this process toward other processes
- **Terminal or TTY**: Terminal to which the process is connected
- **RUID/EUID**: The real/effective user ID, which belongs to the owner of the process
- **RGID/EGID**: The real/effective group owner, the group owner of a process

To see a list of the active processes, there's the `ps` (process status) command, which shows the current list of running processes for the active user:

```
> ps -f
UID     PID  PPID  C  STIME  TTY    TIME      CMD
user    8    4     0  Nov03  pts/0  00:00:00  bash -l -i
user    43   8     0  08:53  pts/0  00:00:00  ps -f
```

Process life cycle

The creation of a new process can happen in two different ways:

- Using a `fork`: This duplicates the calling process. The child (new process) is an exact copy (memory) of the parent (calling process), except for the following:
    - PIDs are different.
    - The PPID of the child equals the PID of the parent.
    - The child does not inherit the following from the parent:
        - Memory locks
        - Semaphore adjustments
        - Outstanding asynchronous I/O operations
        - Asynchronous I/O contexts
- Using an `exec`: This replaces the current process image with a new one, loading the program into the current process space and running it from its entry point.

Foreground and background

When a process is launched, it is normally in the **foreground**, which will prevent communication with the shell until the job is finished or interrupted. Launching a process with an & symbol at the end of the command (cat file.txt &) will launch it in the **background**, making it possible to keep using the shell. The SIGTSTP signal can be sent with *Ctrl + Z*, which allows the user to suspend the foreground process from the shell. It can be resumed with the fg command, or in the background with the bg command.

The jobs command reports the jobs running and their numbers. In the output, the numbers in square brackets are the job numbers that are used by the process control commands, such as fg and bg.

Killing a job

The foreground process can be terminated with the SIGINT signal using *Ctrl + Z*. In order to kill a background process, or send any signal to the process, the kill command can be used.

The kill command receives an argument that can be either of the following:

- The signal sent to the process
- The PID or the job number (with a % prefix)

The more notable signals used are as follows:

- SIGINT: Indicates a termination caused by user input and can be sent by kill command with the -2 value
- SIGTERM: Represents a general purpose termination request not generated by a user as well as a default signal for the kill command with a -6 value
- SIGKILL: A termination handled directly by the operating system that kills the process immediately and has a -9 value

Users, groups, and permissions

Users and groups, together with permissions, are the main entities that are used in Unix operating systems to control access to resources.

Users and groups

Authorization to files and other resources are provided by users and groups. Users have unique usernames that are human-friendly identifiers, but from the operating system side, each user is represent by a unique positive integer: the **User ID (UID)**. Groups are the other authorization mechanism and, as users, they have a name and a **Group ID (GID)**. In the operating system, each process is associated with a user and each file and directory belongs to both a user and a group.

The `/etc/passwd` file contains all this information and more:

```
# username : encrypted password : UID : GID : full name : home directory :
login shell
root:x:0:0:root:/root:/bin/bash
daemon:x:1:1:daemon:/usr/sbin:/usr/sbin/nologin
bin:x:2:2:bin:/bin:/usr/sbin/nologin
sys:x:3:3:sys:/dev:/usr/sbin/nologin
...
user:x:1000:1000:"User Name":/home/user:/bin/bash
```

Users don't use UID directly; they use a combination of username and password in order to start their first process, the interactive shell. Children of the first shell inherit their UID from it, so they keep belonging to the same user.

The UID 0 is reserved for a user known as root, which has special privileges and can do almost anything on the system, such as read/write/execute any file, kill any process, and change running process UIDs.

A group is a logical collection of users, used to share files and directories between them. Each group is independent of other groups and there is no specific relationship between them. For a list of the groups that the current user belongs to, there's the `groups` command. To change group ownership of a file, there's `chgrp`.

Owner, group, and others

Unix files belong to a user and a group. This creates three levels of authorization hierarchy:

- **Owner**: The UID associated with the file
- **Group**: UIDS that belong to the GID associated with the file
- **Others**: Everyone else

Different permissions can be specified for each of these groups, and these permissions are usually decreasing from owner to others. It does not make sense that the owner of a file has less permission on it than its own group or the users outside that group.

Read, write, and execute

Users and groups are used as the first two layers of protection for accessing a file. The user that owns a file has a set of permissions that differs from the file group. Whoever is not the owner and does not belong to the group has different permissions. These three sets of permissions are known as **owner**, **group**, and **other**.

For each element of the set there are three actions that may be carried out: reading, writing, and executing. This is very straightforward for files but it means something different for directories. Read makes it possible to list the contents, write is used to add new links inside, and execute is used to navigate it.

Three permissions are represented by an octal value, where the first bit is read permission, the second is write, and the third is execute. They can also be represented by the letters r, w, and x in sequence and the possible values are as follows:

- 0 or ---: No permissions
- 1 or --x: Execute permission (execute a file or navigate to a directory)
- 2 or -w-: Write permission (write a file or add new files in a directory)
- 3 or -wx: Write and execute
- 4 or r--: Read permission (read a file or list the contents of a directory)
- 5 or r-x: Read and execute
- 6 or rw-: Read and write
- 7 or rwx: Read, write, and execute

The sequence of three octal values represent permission for user, group, and others:

- 777: Everyone can read, write, and execute.
- 700: Owner can read, write, and execute.
- 664: Owner and group can read and write.
- 640: Owner can read and write and group can read.
- 755: Owner can read, write, and execute, while group and others can read and execute.

The ls command with the -l flag (or its alias, ll) shows the list of files and the folder for the current directory with their permissions.

Changing permission

The chmod command makes it possible to change permission on a file or directory. This can be used to override current permissions or to modify them:

- In order to replace permissions, the chmod xxx file command must be issued. *xxx* can be the three octal values representing the permission for the respective tiers or a string that specifies permissions, such as u=rwx, g=rx, or o=r.
- To add or remove one or more permissions, chmod +x file or chmod -x file can be used.

For more information, use the chmod command with the help flag (chmod --help).

Process communications

The operating system is responsible for communication between processes and has different mechanisms to exchange information. These processes are unidirectional, such as exit codes, signals, and pipes, or bidirectional, such as sockets.

Exit codes

Applications communicate their result to the operating system by returning a value called **exit status**. This is an integer value passed to the parent process when the process ends. A list of common exit codes can be found in the /usr/include/sysexits.h file, as shown here:

```
#define EX_OK 0 /* successful termination */
#define EX__BASE 64 /* base value for error messages */
#define EX_USAGE 64 /* command line usage error */
#define EX_DATAERR 65 /* data format error */
#define EX_NOINPUT 66 /* cannot open input */
#define EX_NOUSER 67 /* addressee unknown */
#define EX_NOHOST 68 /* host name unknown */
#define EX_UNAVAILABLE 69 /* service unavailable */
#define EX_SOFTWARE 70 /* internal software error */
#define EX_OSERR 71 /* system error (e.g., can't fork) */
```

```
#define EX_OSFILE 72 /* critical OS file missing */
#define EX_CANTCREAT 73 /* can't create (user) output file */
#define EX_IOERR 74 /* input/output error */
#define EX_TEMPFAIL 75 /* temp failure; user is invited to retry */
#define EX_PROTOCOL 76 /* remote error in protocol */
#define EX_NOPERM 77 /* permission denied */
#define EX_CONFIG 78 /* configuration error */
#define EX__MAX 78 /* maximum listed value */
```

The source for this is as follows: `https://elixir.bootlin.com/linux/latest/source/fs/ext4/ext4.h#L212`.

The exit code of the last command is stored in the `$?` variable, and it can be tested in order to control the flow of the operations. A commonly used operator is `&&` (double ampersand), which executes the next instruction only if the exit code of the first one is 0, such as `stat file && echo something >> file`, which appends something to a file only if it exists.

Signals

Exit codes connect processes and their parents, but signals make it possible to interface any process with another, including itself. They are also asynchronous and unidirectional, but they represent communication from the outside of a process.

The most common signal is `SIGINT`, which tells the application to terminate, and can be sent to a foreground process in a shell with the *Ctrl + C* key combination. However, there are many more options, as shown in the following table:

Name	Number	Description
SIGHUP	1	Controlling terminal is closed
SIGINT	2	Interrupt signal (*Ctrl + C*)
SIGQUIT	3	Quit signal (*Ctrl + D*)
SIGFPE	8	Illegal mathematical operation is attempted
SIGKILL	9	Quits the application immediately
SIGALRM	14	Alarm clock signal

The `kill` command allows you to send a signal to any application, and a comprehensive list of available signals can be shown with the `-l` flag:

```
$ kill -l
 1) SIGHUP       2) SIGINT       3) SIGQUIT      4) SIGILL       5) SIGTRAP
 6) SIGABRT      7) SIGEMT       8) SIGFPE       9) SIGKILL     10) SIGBUS
11) SIGSEGV     12) SIGSYS      13) SIGPIPE     14) SIGALRM     15) SIGTERM
16) SIGURG      17) SIGSTOP     18) SIGTSTP     19) SIGCONT     20) SIGCHLD
21) SIGTTIN     22) SIGTTOU     23) SIGIO       24) SIGXCPU     25) SIGXFSZ
26) SIGVTALRM   27) SIGPROF     28) SIGWINCH    29) SIGPWR      30) SIGUSR1
31) SIGUSR2     32) SIGRTMIN    33) SIGRTMIN+1  34) SIGRTMIN+2  35) SIGRTMIN+3
36) SIGRTMIN+4  37) SIGRTMIN+5  38) SIGRTMIN+6  39) SIGRTMIN+7  40) SIGRTMIN+8
41) SIGRTMIN+9  42) SIGRTMIN+10 43) SIGRTMIN+11 44) SIGRTMIN+12 45) SIGRTMIN+13
46) SIGRTMIN+14 47) SIGRTMIN+15 48) SIGRTMIN+16 49) SIGRTMAX-15 50) SIGRTMAX-14
51) SIGRTMAX-13 52) SIGRTMAX-12 53) SIGRTMAX-11 54) SIGRTMAX-10 55) SIGRTMAX-9
56) SIGRTMAX-8  57) SIGRTMAX-7  58) SIGRTMAX-6  59) SIGRTMAX-5  60) SIGRTMAX-4
61) SIGRTMAX-3  62) SIGRTMAX-2  63) SIGRTMAX-1  64) SIGRTMAX
```

Pipes

Pipes are the last unidirectional communication method between processes. As the name suggests, pipes connect two ends – a process input with another process output – making it possible to process on the same host to communicate in order to exchange data.

These are classified as anonymous or named:

- Anonymous pipes link one process standard output to another process standard input. It can be easily done inside a shell with the `|` operator, linking the output of the command before the pipe as input for the one after the pipe. `ls -l | grep "user"` gets the output of the `ls` command and uses it as input for `grep`.

- Named pipes use a specific file in order to execute the redirect. The output can be redirected to a file with the > (greater) operator, while the < (less) sign allows you to use a file as input for another process. `ls -l > file.txt` saves the output of the command to a file. `cat < file.txt` sends the contents of the file to the command's standard input, and the standard input copies them to the standard output.

It is also possible to append content to a named pipe using the >> (double greater) operator, which will start writing from the end of the file.

Sockets

Unix domain sockets are a bidirectional communication method between applications on the same machine. They are a logical endpoint that is handled by the kernel and manages the data exchange.

The nature of sockets permits using them as stream-oriented, or datagram-oriented. Stream-oriented protocols ensure that messages are delivered before moving to the next chunk of data in order to preserve message integrity. In contrast, message-oriented protocols ignore the data that is not received and keeps sending the following messages, making it a faster but less reliable protocol with very low latency.

The sockets are classified as follows:

- `SOCK_STREAM`: Connection-oriented, ordered, and reliable transmission of a stream of data
- `SOCK_SEQPACKET`: Connection-oriented, ordered, and reliable transmission of message data that has record boundaries
- `SOCK_DGRAM`: Unordered and unreliable transmission of messages

Summary

This chapter provided a general overview of the main Unix components and how they interact with each other. We started with memory management and how it works in Unix, understanding concepts such as **pagination** and **swap.**

Then we analyzed the filesystem, taking a look at the support from modern operating systems, and explained the difference between the existing file types: files, directories, and hard and soft links.

After learning about the concept of inode, we took a look at the structure of a directory in a Unix operating system and explained how to navigate and interact with the filesystem, as well as how to mount and unmount other partitions.

We moved on to processes, running applications in Unix, and their structure and attributes. We analyzed process life cycle, from its creation through `fork` or `exec`, to its end or termination with the `kill` command.

Another important topic was users, groups, and permissions. We saw what a user is, what groups are, how to join them, and how these concepts are used to divide permissions into three groups: user, group, and others. This helped to better understand the Unix permission model, as well as how to change permissions for files and directories.

Finally, we saw how communication between processes works with one-way channels such as signals and exit codes, or bidirectional communication such as sockets.

In the next chapter, we will have a quick overview of the Go language.

Questions

1. Which filesystem is used by modern operating systems?
2. What is an inode? What is inode 0 in Unix?
3. What's the difference between PID and PPID?
4. How do you terminate a process running in the background?
5. What is the difference between user and group?
6. What's the scope of the Unix permission model?
7. Can you explain the difference between signals and exit codes?
8. What's a swap file?

An Overview of Go 3

This chapter will provide an overview of the Go language and its basic functionality. We will provide a short explanation of the language and its features, which we will elaborate on in more detail in the following chapters. This will help us to understand Go better while we're using all its features and applications.

The following topics will be covered in this chapter:

- Features of the language
- Packages and imports
- Basic types, interfaces, and user-defined types
- Variables and functions
- Flow control
- Built-in functions
- Concurrency model
- Memory management

Technical requirements

From this chapter onward, you will need Go installed on your machine. Follow these steps to do this:

1. Download the latest version of Go from `https://golang.org/dl/`.
2. Extract it with `tar -C /usr/local -xzf go$VERSION.$OS-$ARCH.tar.gz`.
3. Add it to `PATH` with `export PATH=$PATH:/usr/local/go/bin`.
4. Ensure that Go is installed with `go version`.

5. Add the export statement in your `.profile` to add it automatically.

6. You can also change the `GOPATH` variable (the default one is `~/go`) if you want to use a different directory for your code.

I also recommend installing Visual Studio Code (`https://code.visualstudio.com/`) with its vscode-go (`https://github.com/Microsoft/vscode-go`) extension, which contains a helper that will install all the tools that are needed to improve the Go development experience.

Language features

Go is a modern server language with great concurrency primitives and a memory system that is mostly automated. It is considered by some to be a successor to C, and it manages to do so in many scenarios because it's performant, has an extensive standard library, and has a great community that provides many third-party libraries which cover, extend, and improve its functionalities.

History of Go

Go was created in 2007 in order to try and address Google's engineering problems, and was publicly announced in 2009, reaching version 1.0 in 2012. The major version is still the same (version 1), while the minor version grows (version 1.1, 1.2, and so on) together with its functionalities. This is done to keep Go's promise of compatibility for all major versions. A draft for two new features (generics and error handling), which will probably be included in version 2.0, were presented in 2018.

The minds behind Go are as follows:

- **Robert Griesemer**: A Google researcher who worked on many projects, including V8 JavaScript engine and design, and the implementation of Sawzall.
- **Rob Pike**: A member of the Unix team, the Plan 9 and Inferno OS development team, and the Limbo programming language design team.
- **Ken Thompson**: A pioneer of computer science, the designer of the original Unix, and inventor of B (direct predecessor of C). Ken was also one of the creators and early developers of the Plan 9 operating system.

Strengths and weaknesses

Go is a very opinionated language; some people love it while some people hate it, mostly due to some of its design choices. Some of the features that have not been well received are as follows:

- Verbose error handling
- Lack of generics
- Missing dependency and version management

The first two points will be addressed with the next major version, while the latter was addressed by the community first (godep, glide, and govendor) and by Google itself with dep for the dependencies, and by gopkg.in (`http://labix.org/gopkg.in`) in terms of version management.

The strengths of the language are countless:

- It's a statically typed language, with all the advantages that this brings, like static type checking.
- It does not require an **integrated development environment (IDE)**, even if it supports many of them.
- The standard library is really impressive, and it could be the only dependency of many projects.
- It has concurrency primitives (channels and goroutines), which hides the hardest parts of writing asynchronous code that are both efficient and safe.
- It comes with a formatting tool, `gofmt`, that unifies the format of Go code, making other people's code look really familiar.
- It produces binaries, with no dependencies, making deployments fast and easy.
- It's minimalistic and has a few keywords, and the code is really easy to read and understand.
- It's duck typed, with implicit interface definition (*if it walks like a duck, swims like a duck, and quacks like a duck, then it probably is a duck*). This comes in very handy when testing a specific function of a system because it can be mocked.
- It is cross platform, meaning that it's able to produce binaries for architecture and OS that are different from the hosting ones.
- There are a huge amount of third-party packages and so it leaves very little behind in terms of functionality. Each package that's hosted on a public repository is indexed and searchable.

Namespace

Now, let's see how Go code is organized. The GOPATH environment variable determines where the code resides. There are three subdirectories inside this:

- src contains all the source code.
- pkg contains compiled packages, which are divided into architecture/OS.
- bin contains the compiled binaries.

The path under the source folder corresponds to the name of the package ($GOPATH/src/my/package/name would be my/package/name).

The go get command makes it possible to fetch and compile packages using it. Go get calls to http://package_name?go-get=1 and if it finds the go-import meta tag, it uses this to fetch the package. The tag should contain the package name, the VCS that was used, and the repository URL, all separated by a space. Let's take a look at an example:

```
<meta name="go-import" content="package-name vcs repository-url">
```

After go get downloads a package, it tries to do the same for the other packages it can't resolve recursively until all the necessary source code is available.

Each file starts with a package definition, that is, package package_name, that needs to be the same for all the files in a directory. If the package produces a binary, the package is main.

Imports and exporting symbols

The package declaration is followed by a series of import statements that specify the packages that are needed.

Importing packages that are not used (unless they are ignored) is a compilation error, which is why the Go formatting tool, gofmt, removes unused packages. There are experimental or community tools like goimports (https://godoc.org/golang.org/x/tools/cmd/goimports) or goreturns (https://github.com/sqs/goreturns) that also add missing imports to a Go file. It's important to avoid having circular dependencies because they will not compile.

Since circular dependencies are not allowed, the packages need to be designed differently to other languages. In order to break the circular dependency, it is good practice to export functionalities from a package or replace the dependency with an interface.

Go reduces all the symbol visibility to a binary model – exported and not exported – unlike many other languages, which have intermediate levels. For each package, all the symbols starting with a capital letter are exported, while everything else is used only inside the package. The exported values can also be used by other packages, while the unexported ones can only be used in the package itself.

An exception is made if one of the package path elements is internal (for example, `my/package/internal/pdf`). This limits itself and its subpackages to be imported only by the nearby packages (for example, `my/package`). This is useful if there are a lot of unexported symbols and you want to break them down into subpackages, while preventing other packages from using it, making basically private subpackages. Take a look at the following list of internal packages:

- `my/package/internal`
- `my/package/internal/numbers`
- `my/package/internal/strings`

These can only be used by `my/package`, and cannot be imported by any other package, including `my`.

The import can have different forms. The standard import form is the complete package name:

```
import "math/rand"
...
rand.Intn
```

A named import replaces the package name with a custom name, which must be used when referencing the package:

```
import r "math/rand"
...
r.Intn
```

The same package import makes the symbol available with no namespace:

```
import . "math/rand"
...
Intn
```

Ignored imports are used to import packages without you having to use them. This makes it possible to execute the `init` function of the package without referencing the package in your code:

```
import _ math/rand
// still executes the rand.init function
```

Type system

The Go type system defines a series of basic types, including bytes, strings and buffers, composite types like slices or maps, and custom types that have been defined by the application.

Basic types

These are Go's basic types:

Category	Types
String	`string`
Boolean	`bool`
Integer	`int`, `int8`, `int16`, `int32`, and `int64`
Unsigned integer	`uint`, `uint8`, `uint16`, `uint32`, and `uint64`
Pointer integer	`uinptr`
Floating pointer	`float32` and `float64`
Complex number	`complex64` and `complex128`

The number of bits of `int`, `uint`, and `uiptr` depends on the architecture (for example, 32 bits for x86, 64 bits for x86_64).

Composite types

Beside the basic types, there are others, which are known as composite types. These are as follows:

Type	Description	Example
Pointer	The address in the memory of a variable	`*int`
Array	A container of the element of the same type with a fixed length	`[2]int`
Slice	Contiguous segment of an array	`[]int`

Map	Dictionary or associative array	`map[int]int`
Struct	A collection of fields that can have different types	`struct{ value int }`
Function	A set of functions with the same parameters and output	`func(int, int) int`
Channel	Type pipes that are used for the communication of elements of the same type	`chan int`
Interface	A specific collection of methods, with an underlying value that supports them	`interface{}`

The empty interface, `interface{}`, is a generic type that can contain any value. Since this interface has no requirements (methods), it can be satisfied by any value.

Interfaces, pointers, slices, functions, channels, and maps can have a void value, which is represented in Go by `nil`:

- Pointers are self-explanatory; they are not referring to any variable address.
- The interface's underlying value can be empty.
- Other pointer types, like slices or channels, can be empty.

Custom-defined types

A package can define its own types by using the `type defined definition` expression, where definition is a type that shares a defined memory representation. Custom types can be defined by basic types:

```
type Message string    // custom string
type Counter int       // custom integer
type Number float32     // custom float
type Success bool       // custom boolean
```

They can also be defined by composite types like slices, maps, or pointers:

```
type StringDuo [2]string           // custom array
type News chan string              // custom channel
type Score map[string]int          // custom map
type IntPtr *int                   // custom pointer
type Transform func(string) string  // custom function
type Result struct {               // custom struct
    A, B int
}
```

They can also be used in combination with other custom types:

```
type Broadcast Message // custom Message
type Timer Counter     // custom Counter
type News chan Message // custom channel of custom type Message
```

The main uses of custom types are to define methods and to make the type specific for a scope, like defining a `string` type called `Message`.

Interface definitions work in a different way. They can be defined by specifying a series of different methods, for instance:

```
type Reader interface {
    Read(p []byte) (n int, err error)
}
```

They can also be a composition of other interfaces:

```
type ReadCloser interface {
    Reader
    Closer
}
```

Alternatively, they can be a combination of the two interfaces:

```
type ReadCloser interface {
    Reader          // composition
    Close() error // method
}
```

Variables and functions

Now that we've looked at types, we'll take a look at how we can instantiate different types in the language. We'll see how variables and constants work first, and then we'll talk about functions and methods.

Handling variables

Variables represent mapping to the content of a portion of contiguous memory. They have a type that defines how much this memory extends, and a value that specifies what's in the memory. Type can be basic, composite, or custom, and its value can be initialized with their zero-value by a declaration, or with another value by assignment.

Declaration

The zero value declaration for a variable is done using the `var` keyword and by specifying the name and type; for example, `var a int`. This could be counter-intuitive for a person that comes from another language, like Java, where type and name order is inverted, but it's actually more human readable.

The `var a int` example describes a variable (`var`) with a name, a, that is an integer (`int`). This kind of expression creates a new variable with a zero-value for the selected type:

Type	Zero value
Numerical types (`int`, `uint`, and `float` types)	0
Strings (`string` types)	" "
Booleans	false
Pointers, interfaces, slices, maps, and channels	nil

The other way of initiating a variable is by assignment, which can have an inferred or specific type. An inferred type is achieved with the following:

- The variable name, followed by the `:=` operator and the value (for example, `a :=` 1), also called **short declaration**.
- The `var` keyword, followed by the name, the `=` operator, and a value (for example, `var a = 1`).

Note that these two methods are almost equivalent, and redundant, but the Go team decided to keep them both for the sake of the compatibility promise of Go 1. The main difference in terms of the short declaration is that the type cannot be specified and it's inferred by the value.

An assignment with a specific type is made with a declaration, followed by an equals sign and the value. This is very useful if the declared type is an interface, or if the type that's inferred is incorrect. Here are some examples:

```
var a = 1           // this will be an int
b := 1              // this is equivalent
var c int64 = 1     // this will be an int64

var d interface{} = 1 // this is declared as an interface{}
```

Some types need to use built-in functions so that they can be initialized correctly:

- `new` makes it possible to create a pointer of a certain type while already allocating some space for the underlying variable.
- `make` initializes slices, maps, and channels:
  - Slices need an extra argument for the size, and an optional one for the capacity of the underlying array.
  - Maps can have one argument for their initial capacity.
  - Channels can also have one argument for its capacity. They are unlike maps, which cannot change.

The initialization of a type using a built-in function is as follows:

```
a := new(int)                      // pointer of a new in variable
sliceEmpty := make([]int, 0)       // slice of int of size 0, and a capacity
of 0
sliceCap := make([]int, 0, 10)     // slice of int of size 0, and a capacity
of 10
map1 = make(map[string]int)        // map with default capacity
map2 = make(map[string]int, 10)    // map with a capacity of 10
ch1 = make(chan int)               // channel with no capacity (unbuffered)
ch2 = make(chan int, 10)           // channel with capacity of 10 (buffered)
```

Operations

We already saw the assignment operation, which gives a new value to a variable using the `=` operator. Let's take a look at few more operators:

- There are the comparison operators, `==` and `!=`, which compare two values and return a Boolean.
- There are some mathematical operations that can be executed on all numerical variables of the same types, that is, `+`, `-`, `*`, and `/`. The sum operation is also used for concatenating strings. `++` and `--` are shorthand for incrementing or decrementing a number by one. `+=`, `-=`, `*=`, and `/=` execute the operation before the equals sign between what's before and what's after the operator and assigns it to the variable on the left. These four operations produce a value of the same type of the variables involved; there are also other comparison operators that are specific to numbers: `<`, `<=`, `>`, and `>=`.
- Some operations are exclusive for integers and produce other integers: `%`, , `&`, `|`, `^`, `&^`, `<<`, and `>>`.

- Others are just for Booleans and produce another Boolean: &&, ||, and !.
- One operator is channel only, <−, and it's used to receive values from or send them to a channel.
- For all non-pointer variables, it is also possible to use &, the reference operator, to obtain the variable address that can be assigned to a pointer variable. The * operator makes it possible to execute a dereference operation on a pointer and obtain the value of the variable indicated by it:

Operator	Name	Description	Example
=	Assignment	Assigns the value to a variable	a = 10
:=	Declaration and assignment	Declares a variables and assigns a value to it	a := 0
==	Equals	Compares two variables, returns a Boolean if they are the same	a == b
!=	Not equals	Compares two variables, returns a Boolean if they are different	a != b
+	Plus	Sum between the same numerical type	a + b
−	Minus	Difference between the same numerical type	a − b
*	Times	Multiplication between the same numerical type	a * b
/	Divided	Division between the same numerical type	a / b
%	Modulo	Remainder after division of the same numerical type	a % b
&	AND	Bit-wise AND	a & b
&^	Bit clear	Bit clear	a &^ b
<<	Left shift	Bit shift to the left	a << b
>>	Right shift	Bit shift to the right	a >> b
&&	AND	Boolean AND	a && b
||	OR	Boolean OR	a || b
!	NOT	Boolean NOT	!a
<−	Receive	Receive from a channel	<−a
−>	Send	Send to a channel	a <− b
&	Reference	Returns the pointer to a variable	&a
*	Dereference	Returns the content of a pointer	*a

Casting

Converting a type into another type is an operation called **casting,** which works slightly differently for interfaces and concrete types:

- Interfaces can be casted to a concrete type that implements it. This conversion can return a second value (a Boolean) and show whether the conversion was successful or not. If the Boolean variable is omitted, the application will panic on a failed casting.

- With concrete types, casting can happen between types that have the same memory structure, or it can happen between numerical types:

```
type N [2]int            // User defined type
var n = N{1,2}
var m [2]int = [2]int(N)  // since N is a [2]int this casting is
possible

var a = 3.14             // this is a float64
var b int = int(a)        // numerical types can be casted, in
this case a will be rounded to 3

var i interface{} = "hello"  // a new empty interface that contains
a string
x, ok := i.(int)          // ok will be false
y := i.(int)              // this will panic
z, ok := i.(string)       // ok will be true
```

There's a special type of conditional operator for casting called **type switch** which allows an application to attempt multiple casts at once. The following is an example of using `interface{}` to check out the underlying value:

```
func main() {
    var a interface{} = 10
    switch a.(type) {
    case int:
        fmt.Println("a is an int")
    case string:
        fmt.Println("a is a string")
    }
}
```

Scope

The variables have a scope or visibility that is also connected to its lifetime. This can be one of the following:

- **Package**: The variable is visible in all the packages; if the variable is exported, it is also visible from other packages.
- **Function**: The variable is visible inside the function that declares it.
- **Control**: The variable is visible inside the block in which it's defined.

The visibility goes down, going from package to block. Since blocks can be nested, outer blocks don't have visibility over the variables of inner blocks.

Two variables in the same scope cannot have the same name, but a variable of an inner scope can reuse an identifier. When this happens, the outer variable is not visible in the inner scope – this is called **shadowing,** and it needs to be kept in mind in order to avoid issues that are hard to identify, such as the following:

```
// this exists in the outside block
var err error
// this exists only in this block, shadows the outer err
if err := errors.New("Doh!"); err !=
    fmt.Println(err)          // this not is changing the outer err
}
fmt.Println(err)              // outer err has not been changed
```

Constants

Go doesn't have immutability for its variables, but defines another type of immutable value called constant. This is defined by the `const` keyword (instead of `var`), and they are values that cannot change. These values can be base types and custom types, which are as follows:

- Numeric (integer, `float`)
- Complex
- String
- Boolean

The value that's specified doesn't have a type when it's assigned to a variable. There's an automatic conversion of both numeric types and string-based types, as shown in the following code:

```
const PiApprox = 3.14

var PiInt int = PiApprox // 3, converted to integer
var Pi float64 = PiApprox // is a float

type MyString string

const Greeting = "Hello!"

var s1 string = Greeting    // is a string
var s2 MyString = Greeting // string is converted to MyString
```

Numerical constants are very useful for mathematical operations because they are just regular numbers, so they can be used with any numerical variable type.

Functions and methods

Functions in Go are identified by the `func` keyword, followed by an identifier, eventual arguments, and return values. Functions in Go can return more than one value at a time. The combination of arguments and returned types is referred to as a **signature**, as shown in the following code:

```
func simpleFunc()
func funcReturn() (a, b int)
func funcArgs(a, b int)
func funcArgsReturns(a, b int) error
```

The part between brackets is the function body, and the `return` statement can be used inside it for an early interruption of the function. If the function returns values, then the return statement must return values of the same type.

The `return` values can be named in the signature; they are zero value variables and if the `return` statement does not specify other values, these values are the ones that are returned:

```
func foo(a int) int {        // no variable for returned type
    if a > 100 {
        return 100
    }
    return a
}
```

```
func bar(a int) (b int) {     // variable for returned type
    if a > 100 {
        b = 100
        return                // same as return b
    }
    return a
}
```

Functions are first-class types in Go and they can also be assigned to variables, with each signature representing a different type. They can also be anonymous; in this case, they are called **closures**. Once a variable is initialized with a function, the same variable can be reassigned with another function with the same signature. Here's an example of assigning a closure to a variable:

```
var a = func(item string) error {
    if item != "elixir" {
        return errors.New("Gimme elixir!")
    }
    return nil
}
```

The functions that are declared by an interface are referred to as methods and they can be implemented by custom types. The method implementation looks like a function, with the exception being that the name is preceded by a single parameter of the implementing type. This is just syntactic sugar – the method definition creates a function under the hood, which takes an extra parameter, that is, the type that implements the method.

This syntax makes it possible to define the same method for different types, each of which will act as a namespace for the function declaration. In this way, it is possible to call a method in two different ways, as shown in the following code:

```
type A int

func (a A) Foo() {}

func main() {
    A{}.Foo()  // Call the method on an instance of the type
    A.Foo(A{}) // Call the method on the type and passing an instance as
argument
}
```

It's important to note that a type and its pointer share the same namespace, so the same method can be implemented just for one of them. The same method cannot be defined for both the type and for its pointer, since declaring the method twice (for a type and its pointer) will produce a compile error (method redeclared). Methods cannot be defined for interfaces, only for concrete types, but interfaces can be used in composite types, including function parameters and return values, as we can see in the following examples:

```
// use error interface with chan
type ErrChan chan error
// use error interface in a map
type Result map[string]error

type Printer interface{
    Print()
}
// use the interface as argument
func CallPrint(p Printer) {
    p.Print()
}
```

The built-in package already defines an interface that is used all over the standard library and in all the packages that are available online – the `error` interface:

```
type error interface {
    Error() string
}
```

This means that whatever type has an `Error()` `string` method can be used as an error, and each package can define its error types according to its needs. This can be used to concisely carry information about the error. In this example, we are defining `ErrKey`, which is specifying that a `string` key was not found. We don't need anything else besides the key to represent our error, as shown in the following code:

```
type ErrKey string

func (e Errkey) Error() string {
    return fmt.Errorf("key %q not found", e)
}
```

Values and pointers

In Go, everything is passed by a value, so when a function or method is invoked, a copy of the variable is made in the stack. This implies that changes that are made to the value are not reflected outside of the called function. Even slices, maps, and other reference types are passed by value, but since their internal structure contains pointers, they act as if they were passed by reference. If a method is defined for a type, it cannot be defined for its pointer and vice versa. The following example has been used to check that the value is updated only inside the method, and that the change does not reflect the `main` function:

```
package main

import (
    "fmt"
)

type A int

func (a A) Foo() {
    a++
    fmt.Println("foo", a)
}

func main() {
    var a A
    fmt.Println("before", a) // 0
    a.Foo() // 1
    fmt.Println("after", a) // 0
}
```

In order to change the original variable, the argument must be a pointer to the variable itself – the pointer will be copied, but it will reference the same memory area, making it possible to change its value. Note that assigning another value pointer, instead of its content, doesn't change what the original pointer refers to because it is a copy.

If we use a method for the type instead of its pointer, we won't see the changes being propagated outside the method.

In the following example, we are using a value receiver. This makes the `User` value in the `Birthday` method a copy of the `User` value in `main`:

```
type User struct {
    Name string
    Age int
}
```

```go
func (u User) Birthday() {
    u.Age++
    fmt.Println(u.Name, "turns", u.Age)
}

func main() {
    u := User{Name: "Pietro", Age: 30}
    fmt.Println(u.Name, "is now", u.Age)
    u.Birthday()
    fmt.Println(u.Name, "is now", u.Age)
}
```

The full example is available at `https://play.golang.org/p/hnUldHLkFJY`.

Since the change is applied to the copy, the original value is left intact, as we can see from the second print statement. If we want to change the value in the original object, we have to use the pointer receiver so that the one that's been copied will be the pointer and the changes will be made to the underlying value:

```go
func (u *User) Birthday() {
    u.Age++
    fmt.Println(u.Name, "turns", u.Age)
}
```

The full example is available at `https://play.golang.org/p/JvnaQL9R7U5`.

We can see that using the pointer receiver allows us to change the underlying value, and that we can change one field of the `struct` or replace the whole `struct` itself, as shown in the following code:

```go
func (u *User) Birthday() {
    *u = User{Name: u.Name, Age: u.Age + 1}
    fmt.Println(u.Name, "turns", u.Age)
}
```

The full example is available at `https://play.golang.org/p/3ugBEZqAood`.

If we try to change the value of the pointer instead of the underlying one, we will edit a new object that is not related to the one that was created in `main`, and the changes will not propagate:

```go
func (u *User) Birthday() {
    u = &User{Name: u.Name, Age: u.Age + 1}
    fmt.Println(u.Name, "turns", u.Age)
}
```

The full example is available at `https://play.golang.org/p/m8u2clKTqEU`.

Some types in Go are automatically passed by reference. This happens because these types are defined internally as structs that contain a pointer. This creates a list of the types, along with their internal definition:

Types	Internal definitions

map	`struct {`
	`    m *internalHashtable`
	`}`

slice	`struct {`
	`    array *internalArray`
	`    len int`
	`    cap int`
	`}`

channel	`struct {`
	`    c *internalChannel`
	`}`

Understanding flow control

In order to control the flow of an application, Go offers different tools – some statements like `if`/`else`, `switch`, and `for` are used in sequential scenarios, whereas others like `go` and `select` are used in concurrent ones.

Condition

The `if` statement verifies a binary condition and executes the code inside the `if` block when the condition is `true`. When an `else` block is present, it's executed when the condition is `false`. This statement also allows a short declaration before the condition, separated by a `;`. This condition can be chained with an `else if` statement, as shown in the following code:

```
if r := a%10; r != 0 { // if with short declaration
    if r > 5 {          // if without declaration
        a -= r
    } else if r < 5 {   // else if statement
        a += 10 - r
    }
} else {                // else statement
    a /= 10
}
```

The other condition statement is `switch`. This allows a short declaration, like `if`, and this is followed by an expression. The value of such an expression can be of any type (not just Boolean), and it's evaluated against a series of `case` statements, each one followed by a block of code. The first statement that matches the expression, if the `switch` and `case` condition are equal, has its block executed.

If a `break` statement is present in the case where the block execution is interrupted, but there's a `fallthrough`, the code inside the following `case` block is executed. A special case called `default` can be used to execute its code if no case is satisfied, as shown in the following code:

```go
switch tier {                          // switch statement
case 1:                                // case statement
    fmt.Println("T-shirt")
    if age < 18{
        break                          // exits the switch block
    }
    fallthrough                        // executes the next case
case 2:
    fmt.Println("Mug")
    fallthrough                        // executes the next case
case 3:
    fmt.Println("Sticker pack")
default:                               // executed if no case is satisfied
    fmt.Println("no reward")
}
```

Looping

The `for` statement is the only looping statement in Go. This requires that you specify three expressions, separated by `;`:

- A short declaration or an assignment to an existing variable
- A condition to verify before each iteration
- An operation to execute at the end of the iteration

All of these statements can be optional, and the absence of a condition means that it is always `true`. A `break` statement interrupts the loop's execution, while `continue` skips the current iteration and proceeds with the next one:

```go
for {                    // infinite loop
    if condition {
        break            // exit the loop
```

```
    }
}

for i < 0 {              // loop with condition
    if condition {
        continue         // skip current iteration and execute next
    }
}

for i:=0; i < 10; i++ {  // loop with declaration, condition and operation
}
```

When a combination of `switch` and `for` are nested, the `continue` and `break` statements refer to the inner flow control statement.

An outer loop or condition can be labelled using a `name:` expression, whereas the name is its identifier, and both `loop` and `continue` can be followed by the name in order to specify where to intervene, as shown in the following code:

```
label:
    for i := a; i<a+2; i++ {
        switch i%3 {
        case 0:
            fmt.Println("divisible by 3")
            break label                            // this break the outer
for loop
        default:
            fmt.Println("not divisible by 3")
        }
    }
```

Exploring built-in functions

We already listed some of the built-in functions that are used to initialize some variables, that is, `make` and `new`. Now, let's go over each function and check out what they do:

- `func append(slice []Type, elems ...Type) []Type`: This function appends elements to the end of a slice. If the underlying array is full, it reallocates the content to a bigger slice before appending.
- `func cap(v Type) int`: Returns the number elements of an array, or of the underlying array if the argument is a slice.
- `func close(c chan<- Type)`: Closes a channel.

- `func complex(r, i FloatType) ComplexType`: Given two floating points, this returns a complex number.
- `func copy(dst, src []Type) int`: Copies elements from a slice to another.
- `func delete(m map[Type]Type1, key Type)`: Removes an entry from a map.
- `func imag(c ComplexType) FloatType`: Returns the imaginary part of a complex number.
- `func len(v Type) int`: Returns the length of an array, slice, map, string, or channel.
- `func make(t Type, size ...IntegerType) Type`: Creates a new slice, map, or channel.
- `func new(Type) *Type`: Returns a pointer to a variable of the specified type, initialized with a zero value.
- `func panic(v interface{})`: Stops the execution of the current goroutine and, if it's not intercepted, the program.
- `func print(args ...Type)`: Writes the arguments to the standard error.
- `func println(args ...Type)`: Writes the arguments to the standard error and adds a new line at the end.
- `func real(c ComplexType) FloatType`: Returns the real part of a complex number.
- `func recover() interface{}`: Stops a panic sequence and captures the panic value.

Defer, panic, and recover

A very important keyword that hides a lot of complexity but makes it possible to execute many operations easily is `defer`. This is applied to a function, method, or closure execution and makes the function it precedes execute before the function returns. A common and very useful usage is closing resources. After opening the resource successfully, a deferred close statement will ensure that it is executed independently of the exit point, as shown in the following code:

```
f, err := os.Open("config.txt")
if err != nil {
    return err
}
defer f.Close() // it will be closed anyways

// do operation on f
```

During a function's lifetime, all the deferred statements are added to a list and before exiting, they are executed in reverse order, from the last to the first `defer`.

These statements are executed even when there's a panic, which is why a deferred function with a `recover` call can be used to intercept a panic in the respective goroutine and avoid the panic that would kill the application otherwise. As well as a manual call to the `panic` function, there is a set of operations that will cause a panic, including the following:

- Accessing a negative or non-existent array/slice index (index out of range)
- Dividing an integer by `0`
- Sending to a closed channel
- Dereferencing on a `nil` pointer (`nil` pointer)
- Using a recursive function call that fills the stack (stack overflow)

Panic should be used for errors that are not recoverable, which is why errors are just values in Go. Recovering a panic should be just an attempt to do something with that error before exiting the application. If an unexpected problem occurs, it's because it hasn't been handled correctly or some checks are missing. This represents a serious issue that needs to be dealt with, and the program needs to change, which is why it should be intercepted and dismissed.

Concurrency model

Concurrency is so central to Go that two of its fundamental tools are just keywords – `chan` and `go`. This is a very clever way of hiding the complexity of a well-designed and implemented concurrency model that is easy to use and understand.

Understanding channels and goroutines

Channels are made for communicating, which is why Go's mantra is as follows:

> *"Do not communicate by sharing memory, share memory by communicating."*

A channel is made for sharing data, and it usually connects two or more execution threads in an application, which makes it possible to send and receive data without worrying about data safety. Go has a lightweight implementation of a thread that is managed by the runtime instead of the operating system, and the best way to make them communicate is through the use of channels.

Creating a new goroutine is pretty easy – you just need to use the `go` operator, followed by a function execution. This includes method calls and closures. If the function has any arguments, they will be evaluated before the routine starts. Once it starts, there is no guarantee that changes to variables from an outer scope will be synchronized if you don't use channels:

```
a := myType{}
go doSomething(a)      // function call
go func() {            // closure call
    // ...
}()                    // note that the closure is executed
go a.someMethod()      // method call
```

We already saw how to create a new channel with the `make` function. If a channel is unbuffered (`0` capacity), sending to the channel is a blocking operation that will wait for another goroutine to receive data from the same channel in order to unlock it. The capacity shows how many messages the channel is capable of holding before sending the next becomes a blocking operation:

```
unbuf := make(chan int)      // unbuffered channel
buf := make(chan int, 3)     // channel with size 3
```

In order to send to a channel, we can use the `<-` operator. If the channel is on the left of the operator, it is a send operation, and if it's on the right, it is a receive operation. The value that's received from a channel can be assigned to a variable, as follows:

```
var ch = make(chan int)
go func() {
    b := <-ch            // receive and assign
    fmt.Println(b)
}()
ch <- 10                 // send to channel
```

Closing a channel can be done with the `close()` function. This operation means that no more values can be sent to the channel. This is why it is usually the responsibility of the sender. Sending to a close channel will cause a `panic` anyway, which is why it should be done by the receiver. Also, when receiving from a channel, a second Boolean variable can be specified in the assignment. This will be true if the channel is still open so that the receiver knows when a channel has been closed:

```
var ch = make(chan int)
go func() {
    b, ok := <-ch        // channel open, ok is true
    b, ok = <-ch         // channel closed, ok is false
    b <- ch              // channel close, b will be a zero value
```

```
}()
ch <- 10                    // send to channel
close(ch)                   // close the channel
```

There is a special control statement called `select` which works exactly like `switch`, but with only operations on channels:

```
var ch1 = make(chan int)
var ch2 = make(chan int)
go func() { ch1 <- 10 }
go func() { <-ch2 }
switch {                    // the first operation that completes is selected
case a := <-ch1:
    fmt.Println(a)
case ch2 <- 20:
    fmt.Println(b)
}
```

Understanding memory management

Go is garbage collected; it manages its own memory with a computational cost. Writing an efficient application requires knowledge of its memory model and internals in order to reduce the garbage collector's work and increase general performance.

Stack and heap

Memory is arranged into two main areas – stack and heap. There is a stack for the application entry point function (`main`), and additional stacks are created with each goroutine, which are stored in the heap. The **stack** is, as its name suggests, a memory portion that grows with each function call, and shrinks when the function returns. The **heap** is made of a series of regions of memory that are dynamically allocated, and their lifetime is not defined a priori as the items in the stack; heap space can be allocated and freed at any time.

All the variables that outlive the function where they are defined are stored in the heap, such as a returned pointer. The compiler uses a process called **escape analysis** to check which variables go on the heap. This can be verified with the `go tool compile -m` command.

Variables in the stack come and go with the function's execution. Let's take a look at a practical example of how the stack works:

```
func main() {
    var a, b = 0, 1
    f1(a,b)
    f2(a)
}

func f1(a, b int) {
    c := a + b
    f2(c)
}

func f2(c int) {
    print(c)
}
```

We have the `main` function calling a function called `f1`, which calls another function called `f2`. Then, the same function is called directly by `main`.

When the `main` function starts, the stack grows with the variables that are being used. In memory, this would look something like the following table, where each column represents the pseudo state of the stack, which it represents how the stack changes in time, going from left to right:

`main` invoked	`f1` invoked	`f2` invoked	`f2` return	`f1` returns	`f2` invoked	`f2` returns	`main` returns
`main()`	`main()`	`main()`	`main()`	`main()`	`main()`	`main()`	// empty
`a = 0`	`a = 0`	`a = 0`	`a = 0`	`a = 0`	`a = 0`	`a = 0`	
`b = 1`	`b = 1`	`b = 1`	`b = 1`	`b = 1`	`b = 1`	`b = 1`	
	`f1()`	`f1()`	`f1()`		`f2()`		
	`a = 0`	`a = 0`	`a = 0`		`c = 0`		
	`b = 1`	`b = 1`	`b = 1`				
	`c = 1`	`c = 1`	`c = 1`				
		`f2()`					
		`c = 1`					

When f1 gets called, the stack grows again by copying the a and b variables in the new part and adding the new variable, c. The same thing happens for f2. When f2 returns, the stack shrinks by getting rid of the function and its variables, which is what happens when f1 finishes. When f2 is called directly, it grows again by recycling the same memory part that was used for f1.

The garbage collector is responsible for cleaning up the unreferenced values in the heap, so avoiding storing data in it is a good way of lowering the work of the **garbage collector (GC)**, which causes a slight decrease in performance in the app when the GC is running.

The history of GC in Go

The GC is responsible for freeing the areas of the heap that are not referenced in any stack. This was originally written in C and had a *stop the world* behavior. The program stopped for a fraction of time, freed the memory, and then resumed its flow.

Go 1.4 started the process of converting the runtime, including the garbage collector, into Go. Translating these parts in Go lays the foundations for easier optimization that already begun with version 1.5, where the GC became much faster and could run concurrently with other goroutines.

From that point on, there has been a lot of optimization and improvement of the process, which managed to reduce the GC time by several orders of magnitude.

Building and compiling programs

Now that we have had a quick overview of all the language features and capabilities, we can focus on how to run and build our applications.

Install

In Go, there are different commands to build packages and applications. The first one is go install, followed by a path or a package name, which creates a compiled version of the packages in the pkg directory inside $GOPATH.

All the compiled packages are organized by the operating system and architecture, which are stored in the $GOOS and $GOARCH environment variables. These settings are visible by using the go env command, together with other information, like compilation flags:

```
$ go env
GOARCH="amd64"
...
GOOS="linux"
GOPATH="/home/user/go"
...
GOROOT="/usr/lib/go-1.12"
...
```

For the current architecture and operating system, all the compiled packages will be placed in the $GOOS_$GOARCH subdirectory:

```
$ ls /home/user/go/pkg/
linux_amd64
```

If the package name is main and it contains a main function, the command will produce an executable binary file, which will be stored in $GOPATH/bin. If a package is already installed and the source file didn't change, it will not be compiled again, which will speed up build time significantly after the first compile.

Build

A binary can also be built in a specific location using the go build command. A specific output file can be defined using the -o flag, otherwise it will be built in the working directory, using the package name as a binary name:

```
# building the current package in the working directory
$ go build .

# building the current package in a specific location
$ go build . -o "/usr/bin/mybinary"
```

When executing the go build command, the argument can be one of the following:

- A package as a relative path (like go build . for the current package or go build ../name)
- A package as an absolute path (go build some/package) that will be looked up in $GOPATH
- A specific Go source file (go build main.go)

The latter case allows you to build a file that is outside $GOPATH and will ignore any other source file in the same directory.

Run

There is a third command that works similarly to build, but also runs the binary. It creates the binary using the `build` command using a temporary directory as output, and executes the binary on the fly:

```
$ go run main.go
main output line 1
main output line 2

$
```

Run can be used over build or install when you're making changes to the source code. If the code is the same, it is probably better to build it once and execute it many times.

Summary

In this chapter, we looked at some of the history of Go and its current pros and cons. After understanding the namespace by looking at how the package system and imports work, we explored its type system with basic, composite, and user-defined types.

We focused on variables by looking at how they can be declared and initialized, which operations are allowed between types, how to cast variables to other types, and how to see what the underlying type of interface is. We saw how scope and shadowing work and the difference between constants and variables. After this, we jumped into functions, a first-class type, and how each signature represents a different type. We then understood how methods are basically functions in disguise and attached to a type that allows custom types to satisfy interfaces.

In addition, we learned how to control the application flow using statements like if, for, and switch. We analyzed the differences between the various control statements and the looping statement and viewed what each built-in function does. Then, we saw how basic concurrency works with channels and goroutines. Finally, we got some notion of how Go's internal memory allocation works, some history and performance of its garbage collector, and how to build, install, and run Go binaries.

In the next chapter, we are going to see how to put some of this into practice by interacting with the filesystem.

Questions

1. What's the difference between an exported and an unexported symbol?
2. Why are custom types important?
3. What is the main limit of a short declaration?
4. What is scope and how does it affect variable shadowing?
5. How can you access a method?
6. Explain the difference between a series of `if/else` and `switch`.
7. In a typical use case, who is generally responsible for closing a channel?
8. What is escape analysis?

Section 2: Advanced File I/O Operations

2

This section covers the application's input and output operations, focusing on files/the filesystem and streams in the Unix OS. It covers many topics, including reading from and writing to files, and other I/O operations. Then we explain how much more Go has to offer in terms of I/O with its interfaces and implementations.

This section consists of the following chapters:

Working with the Filesystem 4

This chapter is all about interacting with the Unix filesystem. Here, we will look at everything from the basic read and write operations to more advanced buffered operations, like token scanning and file monitoring.

All of the information for the user or the system is stored as a file in Unix, so in order to interact with the system and user data, we must interact with the filesystem.

In this chapter, we will see that there are different ways of executing read and write operations, and how each one is focused more on simplicity of the code, the memory usage of the application and its performance, as well as the speed of execution.

The following topics will be covered in this chapter:

- File path manipulation
- Reading files
- Writing files
- Other filesystem operations
- Third-party packages

Technical requirements

This chapter requires Go to be installed and your favorite editor to be set up. For more information, please refer to Chapter 3, *An Overview of Go*.

Handling paths

Go offers a series of functions that make it possible to manipulate file paths that are platform-independent and that are contained mainly in the `path/filepath` and `os` packages.

Working directory

Each process has a directory associated with it called the **working directory**, which is usually inherited from the parent process. This makes it possible to specify relative paths – one that doesn't start with the root folder. This will be / on Unix and macOS and c:\ (or any other drive letter) on Windows.

 An absolute/full path starts with the root directory and it represents the same location in a filesystem, that is, /usr/local.

A relative path doesn't start with a root, and the path starts with the current working directory, that is, documents.

The operating system interprets these paths as being relative to the current directory, so their absolute version is a concatenation of the working directory and the relative path. Let's take a look at the following example:

```
user:~ $ cd documents

user:~/documents $ cd ../videos

user:~/videos $
```

Here, the user is in their home folder, ~. The user specifies to change directory to documents, and the cd command automatically adds the working directory as a prefix to it and moves to ~/documents.

Before moving on to the second command, let's introduce two special files that are available in all the directories for all operative systems:

- .: The dot is a reference to the current directory. If it's the first element of the path, it is the process working directory, otherwise it refers to the path element that precedes it (for example, in ~/./documents, . refers to ~).
- ..: The double dot refers to the parent of the current directory if it's the first element of the path, or to the parent of the directory that it precedes if not (for example, in ~/images/../documents, .. refers to the parent of ~/images, ~).

Knowing this, we can easily infer that the second path is first joined in ~/documents/../videos, the parent element, .., gets resolved, and the final path, ~/videos, is obtained.

Getting and setting the working directory

We can use the `func Getwd() (dir string, err error)` function of the `os` package to find out which path represents the current working directory.

Changing the working directory is done with another function of the same package, that is, `func Chdir(dir string) error`, as shown in the following code:

```
wd, err := os.Getwd()
if err != nil {
    fmt.Println(err)
    return
}
fmt.Println("starting dir:", wd)

if err := os.Chdir("/"); err != nil {
    fmt.Println(err)
    return
}

if wd, err = os.Getwd(); err != nil {
    fmt.Println(err)
    return
}
fmt.Println("final dir:", wd)
```

Path manipulation

The `filepath` package contains less than 20 functions, which is a small number compared to the packages of the standard library, and it's used to manipulate paths. Let's take a quick look at these functions:

- `func Abs(path string) (string, error)`: Returns the absolute version of the path that's passed by joining it to the current working directory (if it's not already absolute), and then cleans it.
- `func Base(path string) string`: Gives the last element of the path (base). For instance, `path/to/some/file` returns the file. Note that if the path is empty, this function returns a `.` (dot) path.
- `func Clean(path string) string`: Returns the shortest version of the path by applying a series of defined rules. It does operations like replacing `.` and `..`, or removing trailing separators.
- `func Dir(path string) string`: Gets the path without its last element. This usually returns the parent directory of the element.

- func `EvalSymlinks(path string) (string, error)`: Returns the path after evaluating symbolic links. The path is relative if the provided path is also relative and doesn't contain symbolic links with absolute paths.

- func `Ext(path string) string`: Gets the file extension of the path, the suffix that starts with the final dot of the last element of the path, and it's an empty string if there's no dot (for example, `docs/file.txt` returns `.txt`).

- func `FromSlash(path string) string`: Replaces all / (slashes) found in the path with the operative system path separator. This function does nothing if the OS is Windows, and it executes a replacement under Unix or macOS .

- func `Glob(pattern string) (matches []string, err error)`: Finds all files matching the specified pattern. If there are no matching files, the result is `nil`. It doesn't report eventual errors that occur during path exploration. It shares syntax with `Match`.

- func `HasPrefix(p, prefix string) bool`: This function is deprecated.

- func `IsAbs(path string) bool`: Shows if the path is absolute or not.

- func `Join(elem ...string) string`: Concatenates multiple path elements by joining them with the filepath separator. Note that this also calls `Clean` on the result.

- func `Match(pattern, name string) (matched bool, err error)`: Verifies that the given name matches the pattern, allowing the use of the wild `char` characters * and ?, and groups or sequences of characters using square brackets.

- func `Rel(basepath, targpath string) (string, error)`: Returns the relative path from the base to the target path, or an error if this is not possible. This function calls `Clean` on the result.

- func `Split(path string) (dir, file string)`: Divides the path into two parts using the final trailing slash. The result is usually the parent path and the file name of the input path. If there is no separator, `dir` will be empty and the file will be the path itself.

- func `SplitList(path string) []string`: Returns a list of paths, separating them with the list separator character, which is : in Unix and macOS and ; in Windows.

- func `ToSlash(path string) string`: Operates the opposite substitution that the `FromSlash` function executes, changing each path separator to a /, doing nothing on Unix and macOS, and executing the replacement in Windows.

- func `VolumeName(path string) string`: This does nothing in platforms that aren't Windows. It returns the path component which refers to the volume. This is done for both local paths and network resources.

- `func Walk(root string, walkFn WalkFunc) error`: Starting from the root directory, this function travels recursively through the file tree, executing the walk function for each entry of the tree. If the walk function returns an error, the walk stops and that error is returned. The function is defined as follows:

```
type WalkFunc func(path string, info os.FileInfo, err error) error
```

 Before moving on to the next example, let's introduce an important variable: `os.Args`. This variable contains at least one value, which is the path that invoked the current process. This can be followed by eventual arguments that are specified in the same call.

We want to realize a small application that lists and counts the number of files in a directory. We can use some of the tools we just saw to achieve this.

An example of the list and count files is shown in the following code:

```go
package main

import (
    "fmt"
    "os"
    "path/filepath"
)

func main() {
    if len(os.Args) != 2 { // ensure path is specified
        fmt.Println("Please specify a path.")
        return
    }
    root, err := filepath.Abs(os.Args[1]) // get absolute path
    if err != nil {
        fmt.Println("Cannot get absolute path:", err)
        return
    }
    fmt.Println("Listing files in", root)
    var c struct {
        files int
        dirs int
    }
    filepath.Walk(root, func(path string, info os.FileInfo, err error)
error {
        // walk the tree to count files and folders
        if info.IsDir() {
            c.dirs++
        } else {
            c.files++
```

```
        }
        fmt.Println("-", path)
        return nil
    })
    fmt.Printf("Total: %d files in %d directories", c.files, c.dirs)
}
```

Reading from files

Getting the contents of a file can be done with an auxiliary function in the io/ioutil package, as well as with the ReadFile function, which opens, reads, and closes the file at once. This uses a small buffer (512 bytes) and loads the whole content in memory. This is not a good idea if the file size is very large, unknown, or if the content of the file can be processed one part at a time.

Reading a huge file from disk at once means copying all the file's content into the primary memory, which is a limited resource. This can cause memory shortages, as well as runtime errors. Reading chunks of a file at a time can help read the content of big files without causing huge memory usage. This is because the same part of the memory will be reused when reading the next chunk.

An example of reading all the content at once is shown in the following code:

```
package main

import (
    "fmt"
    "io/ioutil"
    "os"
)

func main() {
    if len(os.Args) != 2 {
        fmt.Println("Please specify a path.")
        return
    }
    b, err := ioutil.ReadFile(os.Args[1])
    if err != nil {
        fmt.Println("Error:", err)
    }
    fmt.Println(string(b))
}
```

Reader interface

For all operations that read from a disk, there's an interface that is paramount:

```
type Reader interface {
    Read(p []byte) (n int, err error)
}
```

Its job is really simple – fill the given slice of bytes with the content that's been read and return the number of bytes that's been read and an error, if one occurs. There is a special error variable that's defined by the io package, called EOF (**End Of File**), which should be returned when there is no more input available.

A reader makes it possible to process data in chunks (the size is determined by the slice), and if the same slice is reused for the operations that follow, the resulting program is consistently more memory efficient because it is using the same limited part of the memory that allocates the slice.

The file structure

The os.File type satisfies the reader interface and is the main actor that's used to interact with file contents. The most common way to obtain an instance for reading purposes is with the os.Open function. It's very important to remember to close a file when you're done using it – this will not be obvious with short-lived programs, but if an application keeps opening files without closing the ones that it's done with, the application will reach the limit of open files imposed by the operating system and start failing the opening operations.

The shell offers a couple of utilities, as follows:

- One to get the limit of open files – ulimit -n
- Another to check how many files are open by a certain process – lsof -p PID

The previous example opens a file just to show its contents to standard output, which it does by loading all its content in memory. This can be easily optimized with the tools we just mentioned. In the following example, we are using a small buffer and printing its content before it gets overridden by the next read, using a small buffer to keep memory usage at a minimum.

An example of using a byte array as a buffer is shown in the following code:

```
func main() {
    if len(os.Args) != 2 {
        fmt.Println("Please specify a file")
        return
    }
    f, err := os.Open(os.Args[1])
    if err != nil {
        fmt.Println("Error:", err)
        return
    }
    defer f.Close() // we ensure close to avoid leaks

    var (
        b = make([]byte, 16)
    )
    for n := 0; err == nil; {
        n, err = f.Read(b)
        if err == nil {
            fmt.Print(string(b[:n])) // only print what's been read
        }
    }
    if err != nil && err != io.EOF { // we expect an EOF
        fmt.Println("\n\nError:", err)
    }
}
```

The reading loop, if everything works as expected, will continue executing read operations until the file content is over. In that case, the reading loop will return an io.EOF error, which shows that there is no more content available.

Using buffers

A **data buffer**, or just a buffer, is a part of memory that is used to store temporary data while it is moved. Byte buffers are implemented in the bytes package, and they are implemented by an underlying slice that is capable of growing every time the amount of data that needs to be stored will not fit.

If new buffers get allocated each time, the old ones will eventually be cleaned up by the GC itself, which is not an optimal solution. It's always better to reuse buffers instead of allocating new ones. This is because they make it possible to reset the slice while keeping the capacity as it is (the array doesn't get cleared or collected by the GC).

A buffer also offers two functions to show its underlying length and capacity. In the following example, we can see how to reuse a buffer with `Buffer.Reset` and how to keep track of its capacity.

An example of buffer reuse and its underlying capacity is shown in the following code:

```
package main

import (
    "bytes"
    "fmt"
)

func main() {
    var b = bytes.NewBuffer(make([]byte, 26))
    var texts = []string{
        `As he came into the window`,
        `It was the sound of a crescendo
He came into her apartment`,
        `He left the bloodstains on the carpet`,
        `She ran underneath the table
He could see she was unable
So she ran into the bedroom
She was struck down, it was her doom`,
    }
    for i := range texts {
        b.Reset()
        b.WriteString(texts[i])
        fmt.Println("Length:", b.Len(), "\tCapacity:", b.Cap())
    }
}
```

Peeking content

In the previous example, we fixed a number of bytes in order to store the content at every read before printing it. Some functionality is offered by the `bufio` package that makes it possible to use an underlying buffer that is not directly controlled by the user, and makes it possible to execute a very important operation named *peek*.

Peeking is the ability to read content without advancing the reader cursor. Here, under the hood, the peeked data is stored in the buffer. Each reading operation checks whether there's data in this buffer and if there is any, that data is returned while removing it from the buffer. This works like a queue (first in, first out).

The possibilities that this simple operation opens are endless, and they all derive from peeking until the desired sequence of data is found, and then the interested chunk is actually read. The most common uses of this operation include the following:

- The buffers keeps reading from the reader until it finds a newline character (read one line at time).
- The same operation is used until a space is found (read one word at a time).

The structure that allows an application to achieve this behavior is `bufio.Scanner`. This makes it possible to define what the splitting function is and has the following type:

```
type SplitFunc func(data []byte, atEOF bool) (advance int, token []byte,
err error)
```

This function stops when an error is returned, otherwise it returns the number of bytes to advance in the content, and eventually a token. The implemented functions in the package are as follows:

- `ScanBytes`: Byte tokens
- `ScanRunes`: Runes tokens
- `ScanWord`: Words tokens
- `ScanLines`: Line tokens

We could implement a file reader that counts the number of lines with just a reader. The resulting program will try to emulate what the Unix `wc -l` command does.

An example of printing a file and counting lines is shown in the following code:

```go
func main() {
    if len(os.Args) != 2 {
        fmt.Println("Please specify a path.")
        return
    }
    f, err := os.Open(os.Args[1])
    if err != nil {
        fmt.Println("Error:", err)
        return
    }
    defer f.Close()
    r := bufio.NewReader(f) // wrapping the reader with a buffered one
    var rowCount int
    for err == nil {
        var b []byte
        for moar := true; err == nil && moar; {
            b, moar, err = r.ReadLine()
```

```
                if err == nil {
                    fmt.Print(string(b))
                }
            }
            // each time moar is false, a line is completely read
            if err == nil {
                fmt.Println()
                rowCount++

            }
        }
        if err != nil && err != io.EOF {
            fmt.Println("\nError:", err)
            return
        }
        fmt.Println("\nRow count:", rowCount)
    }
```

Closer and seeker

There are two other interfaces that are related to readers: io.Closer and io.Seeker:

```
type Closer interface {
        Close() error
}

type Seeker interface {
        Seek(offset int64, whence int) (int64, error)
}
```

These are usually combined with io.Reader, and the resulting interfaces are as follows:

```
type ReadCloser interface {
        Reader
        Closer
}

type ReadSeeker interface {
        Reader
        Seeker
}
```

The `Close` method ensures that the resource gets released and avoids leaks, while the `Seek` method makes it possible to move the cursor of the current object (for example, a `Writer`) to the desired offset from the start/end of the file, or from its current position.

The `os.File` structure implements this method so that it satisfies all the listed interfaces. It is possible to close the file when the operations are concluded, or to move the current cursor around, depending on what you are trying to achieve.

Writing to file

As we have seen for reading, there are different ways to write files, each one with its own flaws and strengths. In the `ioutil` package, for instance, we have another function called `WriteFile` that allows us to execute the whole operation in one line. This includes opening the file, writing its contents, and then closing it.

An example of writing all a file's content at once is shown in the following code:

```go
package main

import (
    "fmt"
    "io/ioutil"
    "os"
)

func main() {
    if len(os.Args) != 3 {
        fmt.Println("Please specify a path and some content")
        return
    }
    // the second argument, the content, needs to be casted to a byte slice
    if err := ioutil.WriteFile(os.Args[1], []byte(os.Args[2]), 0644); err != nil {
        fmt.Println("Error:", err)
    }
}
```

This example writes all the content at once in a single operation. This requires that we allocate all the content in memory using a byte slice. If the content is too large, memory usage can become a problem for the OS, which could kill the process of our application.

If the size of the content isn't very big and the application is short-lived, it's not a problem if the content gets loaded in memory and written with a single operation. This isn't the best practice for long-lived applications, which are executing reads and writes to many different files. They have to allocate all the content in memory, and that memory will be released by the GC at some point – this operation is not cost-free, which means that is has disadvantages regarding memory usage and performance.

Writer interface

The same principle that is valid for reading also applies for writing – there's an interface in the `io` package that determines writing behaviors, as shown in the following code:

```
type Writer interface {
        Write(p []byte) (n int, err error)
}
```

The `io.Writer` interface defines one method that, given a slice of bytes, returns how many of them have been written and/or if there's been any errors. A writer makes it possible to write data one chunk at a time without there being a requirement to have it all at once. The `os.File` struct also happens to be a writer, and can be used in such a fashion.

We can use a slice of bytes as a buffer to write information piece by piece. In the following example, we will try to combine reading from the previous section with writing, using the `io.Seeker` capabilities to reverse its content before writing it.

An example of reversing the contents of a file is shown in the following code:

```
// Let's omit argument check and file opening, we obtain src and dst
cur, err := src.Seek(0, os.SEEK_END) // Let's go to the end of the file
if err != nil {
    fmt.Println("Error:", err)
    return
}
b := make([]byte, 16)
```

After moving to the end of the file and defining a byte buffer, we enter a loop that goes a little backwards in the file, then reads a section of it, as shown in the following code:

```
for step, r, w := int64(16), 0, 0; cur != 0; {
    if cur < step { // ensure cursor is 0 at max
        b, step = b[:cur], cur
    }
    cur = cur - step
    _, err = src.Seek(cur, os.SEEK_SET) // go backwards
```

```
        if err != nil {
            break
        }
        if r, err = src.Read(b); err != nil || r != len(b) {
            if err == nil { // all buffer should be read
                err = fmt.Errorf("read: expected %d bytes, got %d", len(b), r)
            }
            break
        }
```

Then, we reverse the content and write it to the destination, as shown in the following code:

```
        for i, j := 0, len(b)-1; i < j; i, j = i+1, j-1 {
            switch { // Swap (\r\n) so they get back in place
            case b[i] == '\r' && b[i+1] == '\n':
                b[i], b[i+1] = b[i+1], b[i]
            case j != len(b)-1 && b[j-1] == '\r' && b[j] == '\n':
                b[j], b[j-1] = b[j-1], b[j]
            }
            b[i], b[j] = b[j], b[i] // swap bytes
        }
        if w, err = dst.Write(b); err != nil || w != len(b) {
            if err != nil {
                err = fmt.Errorf("write: expected %d bytes, got %d", len(b), w)
            }
        }
    }
    if err != nil && err != io.EOF { // we expect an EOF
        fmt.Println("\n\nError:", err)
    }
```

Buffers and format

In the previous section, we saw how `bytes.Buffer` can be used to store data temporarily and how it handles its own growth by appending the underlying slice. The `fmt` package extensively uses buffers to execute its operations; these aren't the ones in the bytes package for dependency reasons. This approach is inherent to one of Go's proverbs:

> *"A little copy is better than a little dependency."*

If you have to import a package to use one function or type, you should consider just copying the necessary code into your own package. If a package contains much more than what you need, copying allows you to reduce the final size of the binary. You can also customize the code and tailor it to your needs.

Another use of buffers is to compose a message before writing it. Let's write some code so that we can use a buffer to format a list of books:

```go
const grr = "G.R.R. Martin"

type book struct {
    Author, Title string
    Year int
}

func main() {
    dst, err := os.OpenFile("book_list.txt", os.O_CREATE|os.O_WRONLY, 0666)
    if err != nil {
        fmt.Println("Error:", err)
        return
    }
    defer dst.Close()
    bookList := []book{
        {Author: grr, Title: "A Game of Thrones", Year: 1996},
        {Author: grr, Title: "A Clash of Kings", Year: 1998},
        {Author: grr, Title: "A Storm of Swords", Year: 2000},
        {Author: grr, Title: "A Feast for Crows", Year: 2005},
        {Author: grr, Title: "A Dance with Dragons", Year: 2011},
        // if year is omitted it defaulting to zero value
        {Author: grr, Title: "The Winds of Winter"},
        {Author: grr, Title: "A Dream of Spring"},
    }
    b := bytes.NewBuffer(make([]byte, 0, 16))
    for _, v := range bookList {
        // prints a msg formatted with arguments to writer
        fmt.Fprintf(b, "%s - %s", v.Title, v.Author)
        if v.Year > 0 {
            // we do not print the year if it's not there
            fmt.Fprintf(b, " (%d)", v.Year)
        }
        b.WriteRune('\n')
        if _, err := b.WriteTo(dst); true { // copies bytes, drains buffer
            fmt.Println("Error:", err)
            return
        }
    }
}
```

The buffer is used to compose the book description, where the year is omitted if it's not present. This is very efficient when handling bytes and even better if the buffer is reused each time. If the output of this kind of operation should be a string, there is a very similar struct in the `strings` package called `Builder` that has the same write methods but some differences, such as the following:

- The `String()` method uses the `unsafe` package to convert the bytes into a string, instead of copying them.
- It is not permitted to copy a `strings.Builder` and then write to the copy since this causes a `panic`.

Efficient writing

Each time the `os.File` method, that is, `Write`, is executed, this translates to a system call, which is an operation that comes with some overhead. Generally speaking, it's a good idea, to minimize the number of operations by writing more data at once, thus reducing the time that's spent on such calls.

The `bufio.Writer` struct is a writer that wraps another writer, like `os.File`, and executes write operations only when the buffer is full. This makes it possible to execute a forced write with the `Flush` method, which is generally reserved until the end of the writing process. A good pattern of using a buffer would be the following:

```
var w io.WriteCloser
// initialise writer
defer w.Close()
b := bufio.NewWriter(w)
defer b.Flush()
// write operations
```

`defer` statements are executed in reverse order before returning the current function, so the first `Flush` ensures that whatever is still on the buffer gets written, and then `Close` actually closes the file. If the two operations were executed in reverse order, flush would have tried to write a closed file, returning an error, and failed to write the last chunk of information.

File modes

We saw that the `os.OpenFile` function makes it possible to choose how to open a file with the file mode, which is a `uint32` where each bit has a meaning (like Unix files and folder permissions). The `os` package offers a series of values, each one specifying a mode, and the correct way to combine them is with | (bitwise OR).

The following code shows the ones that are available, and have been taken directly from Go's source:

```
// Exactly one of O_RDONLY, O_WRONLY, or O_RDWR must be specified.
O_RDONLY int = syscall.O_RDONLY // open the file read-only.
O_WRONLY int = syscall.O_WRONLY // open the file write-only.
O_RDWR int = syscall.O_RDWR // open the file read-write.
// The remaining values may be or'ed in to control behavior.
O_APPEND int = syscall.O_APPEND // append data to the file when writing.
O_CREATE int = syscall.O_CREAT // create a new file if none exists.
O_EXCL int = syscall.O_EXCL // used with O_CREATE, file must not exist.
O_SYNC int = syscall.O_SYNC // open for synchronous I/O.
O_TRUNC int = syscall.O_TRUNC // if possible, truncate file when opened.
```

The first three represent the operation that's allowed (read, write, or both), and the others are as follows:

- `O_APPEND`: Before each write, the file offset is positioned at the end of the file.
- `O_CREATE`: Makes it possible to create the file if it doesn't exist.
- `O_EXCL`: If this is used with create, it fails if the file already exists (exclusive creation).
- `O_SYNC`: Executes a read/write operation and verifies its competition.
- `O_TRUNC`: If the file exists, its size is truncated to 0.

Other operations

Read and write are not the only operations that can be executed on a file. In the following section, we'll look at how to use them using the `os` package.

Create

In order to create an empty file, we can call a helper function called `Create`, which opens a new file with a `0666` permission and truncates it if it doesn't exist. Alternatively, we can use `OpenFile` with the `O_CREATE|O_TRUNCATE` mode to specify custom permissions, as shown in the following code:

```
package main

import "os"

func main() {
    f, err := os.Create("file.txt")
    if err != nil {
        fmt.Println("Error:", err)
        return
    }
    f.Close()
}
```

Truncate

To truncate the content of a file under a certain dimension, and leave the file untouched if it's smaller, there is the `os.Truncate` method. Its usage is pretty simple, as shown in the following code:

```
package main

import "os"

func main() {
    // let's keep thing under 4kB
    if err := os.Truncate("file.txt", 4096); err != nil {
        fmt.Println("Error:", err)
    }
}
```

Delete

In order to delete a file, there is another simple function, called os.Remove, as shown in the following code:

```
package main

import "os"

func main() {
    if err := os.Remove("file.txt"); err != nil {
        fmt.Println("Error:", err)
    }
}
```

Move

The os.Rename function makes it possible to change a file name and/or its directory. Note that this operation replaces the destination file if it already exists.

The code for changing a file's name or its directory is as follows:

```
import "os"

func main() {
    if err := os.Rename("file.txt", "../file.txt"); err != nil {
        fmt.Println("Error:", err)
    }
}
```

Copy

There's no unique function that makes it possible to copy a file, but this can easily be done with a reader and a writer with the io.Copy function. The following example shows how to use it to copy from one file to another:

```
func CopyFile(from, to string) (int64, error) {
    src, err := os.Open(from)
    if err != nil {
        return 0, err
    }
    defer src.Close()
    dst, err := os.OpenFile(to, os.O_WRONLY|os.O_CREATE, 0644)
    if err != nil {
```

```
        return 0, err
    }
    defer dst.Close()
    return io.Copy(dst, src)
}
```

Stats

The os package offers the FileInfo interface, which returns the metadata of a file, as shown in the following code:

```
type FileInfo interface {
        Name() string // base name of the file
        Size() int64 // length in bytes for regular files; system-dependent
for others
        Mode() FileMode // file mode bits
        ModTime() time.Time // modification time
        IsDir() bool // abbreviation for Mode().IsDir()
        Sys() interface{} // underlying data source (can return nil)
}
```

The os.Stat function returns information about the file with the specified path.

Changing properties

In order to interact with the filesystem and change these properties, three functions are available:

- func Chmod(name string, mode FileMode) error: Changes the permissions of a file
- func Chown(name string, uid, gid int) error: Changes the owner and group of a file
- func Chtimes(name string, atime time.Time, mtime time.Time) error: Changes the access and modification time of a file

Third-party packages

The community offers many packages that accomplish all kinds of tasks. We will take a quick look at some of these in this section.

Virtual filesystems

Files are a struct in Go, a concrete type, and there's no abstraction around them, whereas a file's information is represented by `os.FileInfo`, which is an interface. This is slightly inconsistent, and there have been many attempts to create a full and consistent abstraction on the filesystem, commonly referred to as a *virtual filesystem*.

Two of the most used packages are as follows:

- `vfs`: `github.com/blang/vfs`
- `afero`: `github.com/spf13/afero`

Even if they are developed separately, they both do the same thing – they define an interface with all the methods of `os.File`, and then they define an interface that implements the function that's available in the `os` package, like creating, opening, and deleting files, and so on.

They offer a version based on `os.File` that's implemented using the standard package, but there's also a memory version that uses data structures that emulate a filesystem. This can be very useful for building a test for any package.

Filesystem events

Go has some experimental features in the `golang.org/x/` package that are located under Go's GitHub handler (`https://github.com/golang/`). The `golang.org/x/sys` package is part of this list and includes a subpackage dedicated to Unix system events. This has been used to build a feature that is missing from Go's file functionality and can be really useful – observing a certain path for events on files like creation, deletion, and update.

The two most famous implementations are as follows:

- `notify`: `github.com/rjeczalik/notify`
- `fsnotify`: `github.com/fsnotify/fsnotify`

Both packages expose a function that allows the creation of watchers. Watchers are structures that contain channels that are in charge of delivering file events. They also expose another function that 's responsible for terminating/closing the watchers and underlying channels.

Summary

In this chapter, we looked at an overview of how to execute file operations in Go. In order to locate files, an extensive array of functions are offered by the `filepath` package. These can help you execute all kind of operations, from composing paths to extracting elements from it.

We also looked at how to read an operation using various methods, from the easiest and less memory efficient ones that are located in the `io/ioutil` package to the ones that require an `io.Writer` implementation to read a fixed chunk of bytes. The importance of the ability to peek content, as implemented in the `bufio` package, allows for a whole set of operations like read word or read line, which stop the reading operation when a token is found. There are other interfaces that are satisfied by files that are very useful; for example, `io.Closer` ensures that the resource is released, and `io.Seeker` is used to move the reading cursor around without the need to actually read the file and discard the output.

Writing a slice of bytes to a file can be achieved in different ways – the `io/ioutil` package makes it possible to do so with a function call, while for more complex or more memory-efficient operations, there's the `io.Writer` interface. This makes it possible to write a slice of bytes at a time, and can be used by the `fmt` package to print formatted data. The buffered writing is used to reduce the amount of actual writing on the disk. This is done with a buffer that collects the content, which then transfers it to a disk every time it gets full.

Finally, we saw how to accomplish other file operations on the filesystem (creating, deleting, copying/moving, and changing a file's attributes) and took a look at some of the filesystem-related third-party packages, that is, virtual filesystem abstraction and filesystem events notifications.

The next chapter will be about streams, and will focus on all the instances of readers and writers that are not related to the filesystem.

Questions

1. What's the difference between absolute and relative paths?
2. How do you obtain or change the current working directory?
3. What are the advantages and downfalls of using `ioutil.ReadAll`?
4. Why are buffers important for reading operations?
5. When should you use `ioutil.WriteFile`?
6. Which operations are available when using a buffered reader that allows peeking?
7. When is it better to read content using a byte buffer?
8. How can buffers be used for writing? What's the advantage of using them?

5
Handling Streams

This chapter deals with streams of data, extending input and output interfaces beyond the filesystem, and how to implement custom readers and writers to serve any purpose.

It also focuses on the missing parts of the input and output utilities that combine them in several different ways, with the goal being to have full control of the incoming and outgoing data.

The following topics will be covered in this chapter:

- Streams
- Custom readers
- Custom writers
- Utilities

Technical requirements

This chapter requires Go to be installed and your favorite editor to be set up. For more information, refer to `Chapter 3`, *An Overview of Go*.

Streams

Writers and readers are not just for files; they are interfaces that abstract flows of data in one direction or another. These flows, often referred to as **streams**, are an essential part of most applications.

Input and readers

Incoming streams of data are considered the `io.Reader` interface if the application has no control over the data flow, and will wait for an error to end the process, receiving the `io.EOF` value in the best case scenario, which is a special error that signals that there is no more content to read, or another error otherwise. The other option is that the reader is also capable of terminating the stream. In this case, the correct representation is the `io.ReadCloser` interface.

Besides `os.File`, there are several implementations of readers spread across the standard package.

The bytes reader

The `bytes` package contains a useful structure that treats a slice of bytes as an `io.Reader` interface, and it implements many more I/O interfaces:

- `io.Reader`: This can act as a regular reader
- `io.ReaderAt`: This makes it possible to read from a certain position onward
- `io.WriterTo`: This makes it possible to write the contents with an offset
- `io.Seeker`: This can move the reader's cursor freely
- `io.ByteScanner`: This can execute a read operation for each byte separately
- `io.RuneScanner`: This can do the same with characters that are made of more bytes

The difference between runes and bytes can be clarified by this example, where we have a string made up of one rune, ⌘, which is represented by three bytes, `e28c98`:

```
func main() {
    const a = `⌘`

    fmt.Printf("plain string: %s\n", a)
    fmt.Printf("quoted string: %q\n",a)

    fmt.Printf("hex bytes: ")
    for i := 0; i < len(a); i++ {
        fmt.Printf("%x ", a[i])
    }
    fmt.Printf("\n")
}
```

The full example is available at `https://play.golang.org/p/gVZOufSmlq1`.

There is also `bytes.Buffer`, which adds writing capabilities on top of `bytes.Reader` and makes it possible to access the underlying slice or get the content as a string.

The `Buffer.String` method converts bytes to string, and this type of casting in Go is done by making a copy of the bytes, because strings are immutable. This means that eventual changes to the buffer are made after the copy will not propagate to the string.

The strings reader

The `strings` package contains another structure that is very similar to the `io.Reader` interface, called `strings.Reader`. This works exactly like the first but the underlying value is a string instead of a slice of bytes.

One of the main advantages of using a string instead of the byte reader, when dealing with strings that need to be read, is the avoidance of copying the data when initializing it. This subtle difference helps with both performance and memory usage because it does fewer allocations and requires the **Garbage Collector (GC)** to clean up the copy.

Defining a reader

Any Go application can define a custom implementation of the `io.Reader` interface. A good general rule when implementing interfaces is to accept interfaces and return concrete types, avoiding unnecessary abstraction.

Let's look at a practical example. We want to implement a custom reader that takes the content from another reader and transforms it into uppercase; we could call this `AngryReader`, for instance:

```go
func NewAngryReader(r io.Reader) *AngryReader {
    return &AngryReader{r: r}
}

type AngryReader struct {
    r io.Reader
}

func (a *AngryReader) Read(b []byte) (int, error) {
    n, err := a.r.Read(b)
    for r, i, w := rune(0), 0, 0; i < n; i += w {
        // read a rune
        r, w = utf8.DecodeRune(b[i:])
        // skip if not a letter
        if !unicode.IsLetter(r) {
```

```
            continue
        }
        // uppercase version of the rune
        ru := unicode.ToUpper(r)
        // encode the rune and expect same length
        if wu := utf8.EncodeRune(b[i:], ru); w != wu {
            return n, fmt.Errorf("%c->%c, size mismatch %d->%d", r, ru, w,
wu)
        }
    }
    return n, err
}
```

This is a pretty straightforward example that uses `unicode` and `unicode/utf8` to achieve its goal:

- `utf8.DecodeRune` is used to obtain the first rune and its width is a portion of the slice read
- `unicode.IsLetter` determines whether a rune is a letter
- `unicode.ToUpper` converts the text into uppercase
- `ut8.EncodeLetter` writes the new letter in the necessary bytes
- The letter and its uppercase version should be the same width

The full example is available at `https://play.golang.org/p/PhdSsbzXcbE`.

Output and writers

The reasoning that applies to incoming streams also applies to outgoing ones. We have the `io.Writer` interface, in which the application can only send data, and the `io.WriteCloser` interface, in which it is also able to close the connection.

The bytes writer

We already saw that the `bytes` package offers `Buffer`, which has both reading and writing capabilities. This implements all the methods of the `ByteReader` interface, plus more than one `Writer` interface:

- `io.Writer`: This can act as a regular writer
- `io.WriterAt`: This makes it possible to write from a certain position onward
- `io.ByteWriter`: This makes it possible to write single bytes

`bytes.Buffer` is a very flexible structure considering that it works for both, `Writer` and `ByteWriter` and works best if reused, thanks to the `Reset` and `Truncate` methods. Instead of leaving a used buffer to be recycled by the GC and make a new buffer, it is better to reset the existing one, keeping the underlying array for the buffer and setting the slice length to 0.

In the previous chapter, we saw a good example of buffer usage:

```
bookList := []book{
    {Author: grr, Title: "A Game of Thrones", Year: 1996},
    {Author: grr, Title: "A Clash of Kings", Year: 1998},
    {Author: grr, Title: "A Storm of Swords", Year: 2000},
    {Author: grr, Title: "A Feast for Crows", Year: 2005},
    {Author: grr, Title: "A Dance with Dragons", Year: 2011},
    {Author: grr, Title: "The Winds of Winter"},
    {Author: grr, Title: "A Dream of Spring"},
}
b := bytes.NewBuffer(make([]byte, 0, 16))
for _, v := range bookList {
    // prints a msg formatted with arguments to writer
    fmt.Fprintf(b, "%s - %s", v.Title, v.Author)
    if v.Year > 0 { // we do not print the year if it's not there
        fmt.Fprintf(b, " (%d)", v.Year)
    }
    b.WriteRune('\n')
    if _, err := b.WriteTo(dst); true { // copies bytes, drains buffer
        fmt.Println("Error:", err)
        return
    }
}
```

A buffer is not made for composing string values. For this reason, when the `String` method is called, bytes get converted into strings, which are immutable, unlike slices. The new string created this way is made with a copy of the current slice, and changes to the slice do not touch the string. It's neither a limit nor a feature; it is an attribute that can lead to errors if used incorrectly. Here's an example of the effect of resetting a buffer and using the `String` method:

```
package main

import (
    "bytes"
    "fmt"
)

func main() {
```

```
        b := bytes.NewBuffer(nil)
        b.WriteString("One")
        s1 := b.String()
        b.WriteString("Two")
        s2 := b.String()
        b.Reset()
        b.WriteString("Hey!")      // does not change s1 or s2
        s3 := b.String()
        fmt.Println(s1, s2, s3)  // prints "One OneTwo Hey!"
}
```

The full example is available at `https://play.golang.org/p/zBjGPMC4sfF`

The string writer

A byte buffer executes a copy of the bytes in order to produce a string. This is why, in version 1.10, `strings.Builder` made its debut. It shares all the write-related methods of a buffer and does not allow access to the underlying slice via the `Bytes` method. The only way of obtaining the final string is with the `String` method, which uses the `unsafe` package under the hood to convert the slice to a string without copying the underlying data.

The main consequence of this is that this struct strongly discourages copying—that's because the underlying slice of the copied slice points to the same array, and writing in the copy would influence the other one. The resulting operation would panic:

```
package main

import (
    "strings"
)

func main() {
    b := strings.Builder{}
    b.WriteString("One")
    c := b
    c.WriteString("Hey!") // panic: strings: illegal use of non-zero
Builder copied by value
}
```

Defining a writer

Any custom implementation of any writer can be defined in the application. A very common case is a decorator, which is a writer that wraps another writer and alters or extends what the original writer does. As for the reader, it is a good habit to have a constructor that accepts another writer and possibly wraps it in order to make it compatible with a lot of the standard library structures, such as the following:

- `*os.File`
- `*bytes.Buffer`
- `*strings.Builder`

Let's get a real-world use case—we want to produce some texts with scrambled letters in each word to test when it starts to become unreadable by a human. We will create a configurable writer that will scramble the letters before writing it to the destination writer and we will create a binary that accepts a file and creates its scrambled version. We will use the `math/rand` package to randomize the scrambling.

Let's define our struct and its constructor. This will accept another writer, a random number generator, and a scrambling `chance`:

```
func NewScrambleWriter(w io.Writer, r *rand.Rand, chance float64)
*ScrambleWriter {
    return &ScrambleWriter{w: w, r: r, c: chance}
}

type ScrambleWriter struct {
    w io.Writer
    r *rand.Rand
    c float64
}
```

The `Write` method needs to execute the bytes without letters as they are, and scramble the sequence of letters. It will iterate the runes, using the `ut8.DecodeRune` function we saw earlier, print whatever is not a letter, and stack all the sequences of letters it can find:

```
func (s *ScrambleWriter) Write(b []byte) (n int, err error) {
    var runes = make([]rune, 0, 10)
    for r, i, w := rune(0), 0, 0; i < len(b); i += w {
        r, w = utf8.DecodeRune(b[i:])
        if unicode.IsLetter(r) {
            runes = append(runes, r)
            continue
        }
        v, err := s.shambleWrite(runes, r)
```

```
        if err != nil {
            return n, err
        }
        n += v
        runes = runes[:0]
    }
    if len(runes) != 0 {
        v, err := s.shambleWrite(runes, 0)
        if err != nil {
            return n, err
        }
        n += v
    }
    return
}
```

When the sequence is over, it will be handled by the shambleWrite method, which will effectively execute a shamble and write the shambled runes:

```
func (s *ScrambleWriter) shambleWrite(runes []rune, sep rune) (n int, err error) {
    //scramble after first letter
    for i := 1; i < len(runes)-1; i++ {
        if s.r.Float64() > s.c {
            continue
        }
        j := s.r.Intn(len(runes)-1) + 1
        runes[i], runes[j] = runes[j], runes[i]
    }
    if sep!= 0 {
        runes = append(runes, sep)
    }
    var b = make([]byte, 10)
    for _, r := range runes {
        v, err := s.w.Write(b[:utf8.EncodeRune(b, r)])
        if err != nil {
            return n, err
        }
        n += v
    }
    return
}
```

The full example is available at https://play.golang.org/p/0Xez--6P7nj.

Built-in utilities

There are a number of other functions in the `io` and `io/ioutil` packages that help with managing readers, writers, and more. Knowing all the tools available will help you to avoid writing unnecessary code, and will guide you in using the best tool for the job.

Copying from one stream to another

There are three main functions in the `io` package that make it possible to transfer data from a writer to a reader. This is a very common scenario; you could be writing the contents from a file opened for reading to another file opened for writing, for instance, or draining a buffer and writing its content as standard output.

We already saw how to use the `io.Copy` function on a file to simulate the behavior of the `cp` command in Chapter 4, *Working with the Filesystem*. This behavior can be extended to any sort of reader and writer implementation, from buffers to network connections.

If the writer is also an `io.WriterTo` interface, the copy calls the `WriteTo` method. If not, it executes a series of writes using a buffer of fixed size (32 KB). If the operation ends with the `io.EOF` value, no error is returned. A common case scenario is the `bytes.Buffer` struct, which is capable of writing its content to another writer and will behave accordingly. Alternatively, if the destination is an `io.ReaderFrom` interface, the `ReadFrom` method is executed.

If the interface is a simple `io.Writer` interface, this method uses a temporary buffer that will be cleaned afterwards. To avoid wasting computing power on garbage collection, and maybe reuse the same buffers, there's another function—the `io.CopyBuffer` function. This has an additional argument, and a new buffer gets allocated only if this extra argument is `nil`.

The last function is `io.CopyN`, which works exactly like `io.Copy` but makes it possible to specify a limit to the number of bytes to be written to the extra argument. If the reader is also `io.Seeker`, it can be useful to write partial content—the seeker first moves the cursor to the correct offset, then a certain number of bytes is written.

Let's make an example of copying n bytes at once:

```
func CopyNOffset(dst io.Writer, src io.ReadSeeker, offset, length int64)
(int64, error) {
  if _, err := src.Seek(offset, io.SeekStart); err != nil {
    return 0, err
  }
  return io.CopyN(dst, src, length)
}
```

The full example is available at https://play.golang.org/p/8wCqGXp5mSZ.

Connected readers and writers

The io.Pipe function creates a pair of readers and writers that are connected. This means that whatever is sent to the writer will be received from the reader. Write operations are blocked if there is still data that is hanging from the last one; only when the reader has finished consuming what has been sent will the new operation be concluded.

This is not an important tool for non-concurrent applications, which are more likely to use concurrent tools such as channels, but when the reader and writer are executing on different goroutines, this can be an excellent mechanism for synchronization, as in the following program:

```
pr, pw := io.Pipe()
go func(w io.WriteCloser) {
    for _, s := range []string{"a string", "another string",
        "last one"} {
            fmt.Printf("-> writing %q\n", s)
            fmt.Fprint(w, s)
    }
    w.Close()
}(pw)
var err error
for n, b := 0, make([]byte, 100); err == nil; {
    fmt.Println("<- waiting...")
    n, err = pr.Read(b)
    if err == nil {
        fmt.Printf("<- received %q\n", string(b[:n]))
    }
}
if err != nil && err != io.EOF {
    fmt.Println("error:", err)
}
```

The full example is available at `https://play.golang.org/p/0YpRK25wFw_c`.

Extending readers

When it comes to incoming streams, there are a lot of functions available in the standard library to improve the capabilities of readers. One of the easiest examples is `ioutil.NopCloser`, which takes a reader and returns `io.ReadCloser`, which does nothing. This is useful if a function is in charge of releasing a resource, but the reader used is not `io.Closer` (like in `bytes.Buffer`).

There are two tools that constrain the number of bytes read. The `ReadAtLeast` function defines a minimum number of bytes to read. The result will be `EOF` only if there are no bytes to read; otherwise, if a smaller number of bytes is read before `EOF`, `ErrUnexpectedEOF` will be returned. If the bytes buffer is shorter than the bytes requested, which does not make sense, there will be a `ErrShortBuffer`. In the event of a reading error, the function manages to read at least the desired number of bytes, and that error is dropped.

There is then `ReadFull`, which is expected to fill the buffer and will return `ErrUnexpectedEOF` otherwise.

The other constraining function is `LimitReader`. This function is a decorator that gets a reader and returns another reader that will return `EOF` once the desired bytes are read. This could be used for a preview of the content of an actual reader, as in the following example:

```
s := strings.NewReader(`Lorem Ipsum is simply dummy text of the printing
and typesetting industry. Lorem Ipsum has been the industry's standard
dummy text ever since the 1500s, when an unknown printer took a galley of
type and scrambled it to make a type specimen book. It has survived not
only five centuries, but also the leap into electronic typesetting,
remaining essentially unchanged.`)
    io.Copy(os.Stdout, io.LimitReader(s, 25)) // will print "Lorem Ipsum is
simply dum"
```

The full example is available at `https://play.golang.org/p/LllOdWg9uyU`.

More readers can be combined in a sequence with the `MultiReader` function will read each part sequentially until it reaches `EOF`, and then jump to the next one.

One reader and one writer can be connected so that whatever comes from the reader is copied to the writer—the opposite of what happens with `io.Pipe`. This is done via `io.TeeReader`.

Let's try to use it to create a writer that acts as a search engine in the filesystem, printing only the rows with a match to the query requested. We want a program that does the following:

- Reads a directory path and a string to search from the arguments
- Gets a list of files in the selected path
- Reads each file and passes the lines that contain the selected string to another writer
- This other writer will inject color characters to highlight the string and copy its content to the standard output

Let's start with color injection. In a Unix shell, colored output is obtained with the following sequence:

- \xbb1: An escape character
- [: An opening bracket
- 39: A number
- m: The letter *m*

The number determines both the background and foreground color. For this example, we'll use 31 (red) and 39 (default).

We are creating a writer that will print the rows with a match and highlight the text:

```
type queryWriter struct {
    Query []byte
    io.Writer
}

func (q queryWriter) Write(b []byte) (n int, err error) {
    lines := bytes.Split(b, []byte{'\n'})
    l := len(q.Query)
    for _, b := range lines {
        i := bytes.Index(b, q.Query)
        if i == -1 {
            continue
        }
        for _, s := range [][]byte{
            b[:i], // what's before the match
            []byte("\x1b[31m"), //star red color
            b[i : i+l], // match
            []byte("\x1b[39m"), // default color
            b[i+l:], // whatever is left
        } {
```

```
            v, err := q.Writer.Write(s)
            n += v
            if err != nil {
                return 0, err
            }
        }
        fmt.Fprintln(q.Writer)
    }
    return len(b), nil
}
```

This will be used with `TeeReader` with an open file, so that reading the file will write to `queryWriter`:

```
func main() {
    if len(os.Args) < 3 {
        fmt.Println("Please specify a path and a search string.")
        return
    }
    root, err := filepath.Abs(os.Args[1]) // get absolute path
    if err != nil {
        fmt.Println("Cannot get absolute path:", err)
        return
    }
    q := []byte(strings.Join(os.Args[2:], " "))
    fmt.Printf("Searching for %q in %s...\n", query, root)
    err = filepath.Walk(root, func(path string, info os.FileInfo,
        err error) error {
            if info.IsDir() {
                return nil
            }
            fmt.Println(path)
            f, err := os.Open(path)
            if err != nil {
                return err
            }
        defer f.Close()

        _, err = ioutil.ReadAll(io.TeeReader(f, queryWriter{q, os.Stdout}))
        return err
    })
    if err != nil {
        fmt.Println(err)
    }
}
```

As you can see, there is no need to write; reading from the file automatically writes to the query writer that is connected to the standard output.

Writers and decorators

There are a plethora of tools to enhance, decorate, and use for readers, but the same thing does not apply to writers.

There is also the `io.WriteString` function, which prevents unnecessary conversions from strings to bytes. First, it checks whether the writer supports string writing, attempting a cast to `io.stringWriter`, an unexported interface with just the `WriteString` method, then writes the string if successful, or converts it into bytes otherwise.

There is the `io.MultiWriter` function, which creates a writer that replicates the information to a series of other writers, which it receives upon creation. A practical example is writing some content while showing it on the standard output, as in the following example:

```
r := strings.NewReader("let's read this message\n")
b := bytes.NewBuffer(nil)
w := io.MultiWriter(b, os.Stdout)
io.Copy(w, r) // prints to the standard output
fmt.Println(b.String()) // buffer also contains string now
```

The full example is available at https://play.golang.org/p/ZWDF2vCDfsM.

There is also a useful variable, `ioutil.Discard`, which is a writer that writes to `/dev/null`, a null device. This means that writing to this variable ignores the data.

Summary

In this chapter, we introduced the concept of streams for describing incoming and outgoing flows of data. We saw that the reader interface represents the data received, while the writer is the sent data.

We compared the different readers that are available in the standard package. We looked at files in the previous chapter, and in this one we added byte and string readers to the list. We learned how to implement custom readers with an example, and saw that it's always good to design a reader to be built on top of another.

Then, we focused on writers. We discovered that files are also writers if opened correctly and that there are several writers in the standard package, including the byte buffer and the string builder. We also implemented a custom writer and saw how to handle bytes and runes with the `utf8` package.

Finally, we explored the remaining functionality in `io` and `ioutil`, analyzing the various tools offered for copying data, and connecting readers and writers. We also saw which decorators are available for improving or changing readers' and writers' capabilities.

In the next chapter, we will talk about pseudo terminal applications, and we will use all that knowledge to build some of them.

Questions

1. What's a stream?
2. What interfaces abstract incoming streams?
3. Which interfaces represent outgoing streams?
4. When should a bytes reader be used? When should a string reader be used instead?
5. What's the difference between a string builder and a bytes buffer?
6. Why should readers and writers implementations accept an interface as input?
7. How does a pipe differ from `TeeReader`?

6
Building Pseudo-Terminals

This chapter will introduce pseudo-terminal applications. Many programs (such as SQL or SSH clients) are built as pseudo-terminal because it enables interactive use from inside a terminal. These types of application are very important because they allow us to have control of an application in an environment where there is no graphical interface available, such as when connecting via **Secure Shell (SSH)** to a server. This chapter will guide you through the creation of some applications of this type.

The following topics will be covered in this chapter:

- Terminals and pseudo-terminals
- Basic pseudo-terminals
- Advanced pseudo-terminals

Technical requirements

This chapter requires Go to be installed and your favorite editor to be set up. For more information, you can refer to Chapter 3, *An Overview of Go*.

Understanding pseudo-terminals

Pseudo-terminals, or pseudo teletypes, are applications that run under a Terminal or teletype and emulate its behavior. It's a very convenient way of creating interactive software that is capable of running inside a Terminal without any graphical interface. This is because it uses the Terminal itself to emulate one.

Beginning with teletypes

Teletype (TTY) or teleprinter is the name for typewriters with an electromechanical system that was controlled through a serial port. It was connected to a computer that was capable of sending information to the device to print it. The data was made by a sequence of finite symbols such as ASCII characters, with a fixed font. These devices acted as user interfaces for early computers, so they were—in a sense—precursors to the modern screens.

When screens to replaced printers as output devices, their content was organized in a similar fashion: a two dimensional matrix of characters. In their early stages they were called glass TTY, and the character display was still part of the display itself, controlled by its own logic circuits. With the arrival of the first video display cards, computers were capable of having an interface that was not hardware-dependent.

Text-only consoles used as a main interface for operating systems inherit their name from TTY and are referred to as consoles. Even if the OS runs a graphical environment like on a modern OS, a user can always access a certain number of virtual consoles that work as a **Command-Line Interface (CLI)**, often referred to as a shell.

Pseudo teletypes

Many applications are designed to work inside a shell, but some of them are mimicking the shell's behavior. Graphical interfaces have a Terminal emulator that is designed for executing shells. These types of applications are called **pseudo-teletypes (PTY)**. In order to be considered a PTY, an application needs to be capable of the following:

- Accepting input from the user
- Sending input to the console and receiving the output
- Showing this output to the user

There are some examples already in the available Linux utilities, **screen** being the most notable. It is a pseudo-terminal application that allows the user to use multiple shells and control them. It can open and close new shells, and switch between all opened shells. It allows the user to name a session, so that, if it's killed for any unexpected reason, the user can resume the session.

Creating a basic PTY

We'll start with a simple version of a pseudo-terminal by creating an input manager, then by creating a command selector, and finally by creating the command execution.

Input management

The standard input can be used to receive user commands. We can start by using a buffered input to read lines and print them. In order to read a line, there is a useful command, bufio.Scanner, that already provides a line reader. The code will look similar to the following code snippet:

```
s := bufio.NewScanner(os.Stdin)
w := os.Stdout
fmt.Fprint(w, "Some welcome message\n")
for {
    s.Scan() // get next the token
    fmt.Fprint(w, "You wrote \"")
    w.Write(s.Bytes())
    fmt.Fprintln(w, "\"\n") // writing back the text
}
```

Since this code has no exit point, we can start by creating the first command, exit, that will terminate the shell execution. We can do a small change in the code to make this work, as shown in the following:

```
s := bufio.NewScanner(os.Stdin)
w := os.Stdout
fmt.Fprint(w, "Some welcome message\n")
for {
    s.Scan() // get next the token
    msg := string(s.Bytes())
    if msg == "exit" {
        return
    }
    fmt.Fprintf (w, "You wrote %q\n", msg) // writing back the text
}
```

Now the application has an exit point other than the kill command. For now, it does not implement any command, besides the exit one, and all it does is print back whatever you type.

Selector

In order to be able to interpret the commands correctly, the message needs to be split into arguments. This is the same logic that the operating system applies to argument passed to a process. The `strings.Split` function does the trick, by specifying a space as a second argument and breaking the string into words, as shown in the following code:

```
args := strings.Split(string(s.Bytes()), " ")
cmd := args[0]
args = args[1:]
```

It's possible to execute any sort of check on `cmd`, such as the following `switch` statement:

```
switch cmd {
case "exit":
    return
case "someCommand":
    someCommand(w, args)
case "anotherCommand":
    anotherCommand(w, args)
}
```

This allows the user to add a new command by defining a function and adding a new `case` to the `switch` statement.

Command execution

Now that everything is set, it only remains to define what the various commands will actually do. We can define the type of the function that executes a command and how the switch behaves:

```
var cmdFunc func(w io.Writer, args []string) (exit bool)
switch cmd {
case "exit":
    cmdFunc = exitCmd
}
if cmdFunc == nil {
    fmt.Fprintf(w, "%q not found\n", cmd)
    continue
}
if cmdFunc(w, args) { // execute and exit if true
    return
}
```

The return value tells the application whether it needs to terminate or not and allows us to define our `exit` function easily, without it being a special case:

```go
func exitCmd(w io.Writer, args []string) bool {
    fmt.Fprintf(w, "Goodbye! :)")
    return true
}
```

We could implement any type of command now, depending on the scope of our application. Let's create a `shuffle` command, which will print the arguments in a shuffled order with the help of the `math/rand` package:

```go
func shuffle(w io.Writer, args ...string) bool {
    rand.Shuffle(len(args), func(i, j int) {
        args[i], args[j] = args[j], args[i]
    })
    for i := range args {
        if i > 0 {
            fmt.Fprint(w, " ")
        }
        fmt.Fprintf(w, "%s", args[i])
    }
    fmt.Fprintln(w)
    return false
}
```

We could interact with the filesystem and files by creating a print command that will show in the output of the contents of a file:

```go
func print(w io.Writer, args ...string) bool {
    if len(args) != 1 {
        fmt.Fprintln(w, "Please specify one file!")
        return false
    }
    f, err := os.Open(args[0])
    if err != nil {
        fmt.Fprintf(w, "Cannot open %s: %s\n", args[0], err)
    }
    defer f.Close()
    if _, err := io.Copy(w, f); err != nil {
        fmt.Fprintf(w, "Cannot print %s: %s\n", args[0], err)
    }
    fmt.Fprintln(w)
    return false
}
```

Some refactor

The current version of the pseudo-terminal application could be improved with a little refactoring. We could start by defining a command as a custom type, with a couple of methods that describe its behavior:

```
type cmd struct {
    Name string // the command name
    Help string // a description string
    Action func(w io.Writer, args ...string) bool
}

func (c cmd) Match(s string) bool {
  return c.Name == s
}

func (c cmd) Run(w io.Writer, args ...string) bool {
  return c.Action(w, args...)
}
```

All of the information about each command can be self-contained in a structure. We can also start defining commands that depend on other commands, such as a help command. If we have a slice or a map of commands defined somewhere in the var cmds [] cmd package, the help command will be as follows:

```
help := cmd{
    Name: "help",
    Help: "Shows available commands",
    Action: func(w io.Writer, args ...string) bool {
        fmt.Fprintln(w, "Available commands:")
        for _, c := range cmds {
            fmt.Fprintf(w, " - %-15s %s\n", c.Name, c.Help)
        }
        return false
    },
}
```

The part of the main loop that selects the correct command will be slightly different; it will need to find the match in the slice and execute it:

```
for i := range cmds {
    if !cmds[i].Match(args[0]) {
        continue
    }
    idx = i
    break
}
```

```
if idx == -1 {
    fmt.Fprintf(w, "%q not found. Use `help` for available commands\n",
args[0])
    continue
}
if cmds[idx].Run(w, args[1:]...) {
    fmt.Fprintln(w)
    return
}
```

Now that there is a `help` command that shows the list of available commands, we can advocate using it each time the user specifies a command that does not exist—as we are currently doing when checking whether the index has changed from its default value, -1.

Improving the PTY

Now that we have seen how to create a basic pseudo-terminal, we will see how to improve it with some additional features.

Multiline input

The first thing that can be improved is the relationship between arguments and spacing, by adding support for quoted strings. This can be done with `bufio.Scanner` with a custom split function that behaves like `bufio.ScanWords` apart from the fact that it's aware of quotes. The following code demonstrates this:

```
func ScanArgs(data []byte, atEOF bool) (advance int, token []byte, err
error) {
    // first space
    start, first := 0, rune(0)
    for width := 0; start < len(data); start += width {
        first, width = utf8.DecodeRune(data[start:])
        if !unicode.IsSpace(first) {
            break
        }
    }
    // skip quote
    if isQuote(first) {
        start++
    }
```

The function has a first block that skips spaces and finds the first non-space character; if that character is a quote, it is skipped. Then, it looks for the first character that terminates the argument, which is a space for normal arguments, and the respective quote for the other arguments:

```
// loop until arg end character
for width, i := 0, start; i < len(data); i += width {
    var r rune
    r, width = utf8.DecodeRune(data[i:])
    if ok := isQuote(first); !ok && unicode.IsSpace(r) || ok
        && r == first {
            return i + width, data[start:i], nil
    }
}
```

If the end of file is reached while in a quoted context, the partial string is returned; otherwise, the quote is not skipped and more data is requested:

```
// token from EOF
if atEOF && len(data) > start {
    return len(data), data[start:], nil
}
if isQuote(first) {
    start--
}
return start, nil, nil
}
```

The full example is available at: `https://play.golang.org/p/CodJjcpzlLx`.

Now we can use this as our line for parsing the arguments, while using the helper structure, `argsScanner`, defined as follows:

```
type argsScanner []string

func (a *argsScanner) Reset() { *a = (*a)[0:0] }

func (a *argsScanner) Parse(r io.Reader) (extra string) {
    s := bufio.NewScanner(r)
    s.Split(ScanArgs)
    for s.Scan() {
        *a = append(*a, s.Text())
    }
    if len(*a) == 0 {
        return ""
    }
    lastArg := (*a)[len(*a)-1]
```

```
        if !isQuote(rune(lastArg[0])) {
            return ""
        }
        *a = (*a)[:len(*a)-1]
        return lastArg + "\n"
}
```

This custom slice will allow us to receive lines with quotes and new lines between quotes by changing how the loop works:

```
func main() {
  s := bufio.NewScanner(os.Stdin)
  w := os.Stdout
  a := argsScanner{}
  b := bytes.Buffer{}
  for {
        // prompt message
        a.Reset()
        b.Reset()
        for {
            s.Scan()
            b.Write(s.Bytes())
            extra := a.Parse(&b)
            if extra == "" {
                break
            }
            b.WriteString(extra)
        }
        // a contains the split arguments
    }
}
```

Providing color support to the pseudo-terminal

The pseudo-terminal can be improved by providing colored output. We have already seen that, in Unix, there are escape sequences that can change the background and foreground color. Let's start by defining a custom type:

```
type color int

func (c color) Start(w io.Writer) {
    fmt.Fprintf(w, "\x1b[%dm", c)
}

func (c color) End(w io.Writer) {
    fmt.Fprintf(w, "\x1b[%dm", Reset)
```

```
}

func (c color) Sprintf(w io.Writer, format string, args ...interface{}) {
    c.Start(w)
    fmt.Fprintf(w, format, args...)
    c.End(w)
}

// List of colors
const (
    Reset color = 0
    Red color = 31
    Green color = 32
    Yellow color = 33
    Blue color = 34
    Magenta color = 35
    Cyan color = 36
    White color = 37
)
```

This new type can be used to enhance commands with colored output. For instance, let's take the `shuffle` command and use alternate colors to distinguish between strings, now that we support arguments with spaces:

```
func shuffle(w io.Writer, args ...string) bool {
    rand.Shuffle(len(args), func(i, j int) {
        args[i], args[j] = args[j], args[i]
    })
    for i := range args {
        if i > 0 {
            fmt.Fprint(w, " ")
        }
        var f func(w io.Writer, format string, args ...interface{})
        if i%2 == 0 {
            f = Red.Fprintf
        } else {
            f = Green.Fprintf
        }
        f(w, "%s", args[i])
    }
    fmt.Fprintln(w)
    return false
}
```

Suggesting commands

When the specified command does not exist, we can suggest some similar commands. In order to do so, we can use the Levenshtein distance formula, which measures the similarity between strings by counting deletions, insertions, and substitutions needed to go from one string to the other.

In the following code, we will use the `agnivade/levenshtein` package, which will be obtained through the `go get` command:

```
go get github.com/agnivade/levenshtein/...
```

Then, we define a new function to call when there is no match with existing commands:

```go
func commandNotFound(w io.Writer, cmd string) {
    var list []string
    for _, c := range cmds {
        d := levenshtein.ComputeDistance(c.Name, cmd)
        if d < 3 {
            list = append(list, c.Name)
        }
    }
    fmt.Fprintf(w, "Command %q not found.", cmd)
    if len(list) == 0 {
        return
    }
    fmt.Fprint(w, " Maybe you meant: ")
    for i := range list {
        if i > 0 {
            fmt.Fprint(w, ", ")
        }
        fmt.Fprintf(w, "%s", list[i])
    }
}
```

Extensible commands

The current limitation of our pseudo-terminal is its extensibility. If a new command needs to be added, it needs to be added directly into the main package. We can think of a way of separating the commands from the main package and allowing other users to extend the functionality with their commands:

1. The first step is creating an exported command. Let's use an interface to define a command, so that the user can implement their own:

```
// Command represents a terminal command
type Command interface {
    GetName() string
    GetHelp() string
    Run(input io.Reader, output io.Writer, args ...string) (exit
bool)
}
```

2. Now we can specify a list of commands and a function for other packages to add other commands:

```
// ErrDuplicateCommand is returned when two commands have the same
name
var ErrDuplicateCommand = errors.New("Duplicate command")

var commands []Command

// Register adds the Command to the command list
func Register(command Command) error {
    name := command.GetName()
    for i, c := range commands {
        // unique commands in alphabetical order
        switch strings.Compare(c.GetName(), name) {
        case 0:
            return ErrDuplicateCommand
        case 1:
            commands = append(commands, nil)
            copy(commands[i+1:], commands[i:])
            commands[i] = command
            return nil
        case -1:
            continue
        }
    }
    commands = append(commands, command)
    return nil
}
```

3. We can provide a base implementation of the command, to execute simple functions:

```
// Base is a basic Command that runs a closure
type Base struct {
    Name, Help string
    Action func(input io.Reader, output io.Writer, args ...string)
bool
}

func (b Base) String() string { return b.Name }

// GetName returns the Name
func (b Base) GetName() string { return b.Name }

// GetHelp returns the Help
func (b Base) GetHelp() string { return b.Help }

// Run calls the closure
func (b Base) Run(input io.Reader, output io.Writer, args
...string) bool {
    return b.Action(input, output, args...)
}
```

4. We can provide a function that matches a command to the name:

```
// GetCommand returns the command with the given name
func GetCommand(name string) Command {
    for _, c := range commands {
        if c.GetName() == name {
            return c
        }
    }
    return suggest
}
```

5. We can use the logic from the previous example to make this function return the suggestion command, which will be defined as follows:

```
var suggest = Base{
    Action: func(in io.Reader, w io.Writer, args ...string) bool {
        var list []string
        for _, c := range commands {
            name := c.GetName()
            d := levenshtein.ComputeDistance(name, args[0])
            if d < 3 {
                list = append(list, name)
            }
```

```
    }
    fmt.Fprintf(w, "Command %q not found.", args[0])
    if len(list) == 0 {
        return false
    }
    fmt.Fprint(w, " Maybe you meant: ")
    for i := range list {
        if i > 0 {
            fmt.Fprint(w, ", ")
        }
        fmt.Fprintf(w, "%s", list[i])
    }
    return false
    },
}
```

6. We can already register a couple of commands in the exit and help packages.
   Only help can be defined here, because the list of commands is private:

```
func init() {
    Register(Base{Name: "help", Help: "...", Action: helpAction})
    Register(Base{Name: "exit", Help: "...", Action: exitAction})
}

func helpAction(in io.Reader, w io.Writer, args ...string) bool {
    fmt.Fprintln(w, "Available commands:")
    for _, c := range commands {
        n := c.GetName()
        fmt.Fprintf(w, " - %-15s %s\n", n, c.GetHelp())
    }
    return false
}

func exitAction(in io.Reader, w io.Writer, args ...string) bool {
    fmt.Fprintf(w, "Goodbye! :)\n")
    return true
}
```

This approach will allow a user to use the commandBase struct to create a simple
command, or to embed it or use a custom struct if their command requires it (like a
command with a state):

```
// Embedded unnamed field (inherits method)
type MyCmd struct {
    Base
    MyField string
}
```

```
// custom implementation
type MyImpl struct{}

func (MyImpl) GetName() string { return "myimpl" }
func (MyImpl) GetHelp() string { return "help string"}
func (MyImpl) Run(input io.Reader, output io.Writer, args ...string) bool {
    // do something
    return true
}
```

The difference between the `MyCmd` struct and the `MyImpl` struct is that one can be used as decorator for another command, while the second is a different implementation so it can't interact with another command.

Commands with status

Until now, we have created commands that don't have an internal state. But some commands can keep an internal state and change its behavior accordingly. The state could be limited to the session itself or it could be shared across multiple sessions. The more obvious example is the command history in the Terminal, where all commands executed are stored and retained between sessions.

Volatile status

The easiest thing to implement is a status that is not persistent and gets lost when the application exits. All we need to do is create a custom data structure that hosts the status and satisfies the command interface. The methods will belong to the pointer to the type, as otherwise they will not be able to modify the data.

In the following example, we will create a very basic memory storage that works as a stack (first in, last out) with arguments. Let's start with the push and pop functionalities:

```
type Stack struct {
    data []string
}

func (s *Stack) push(values ...string) {
    s.data = append(s.data, values...)
}

func (s *Stack) pop() (string, bool) {
    if len(s.data) == 0 {
        return "", false
```

```
    }
    v := s.data[len(s.data)-1]
    s.data = s.data[:len(s.data)-1]
    return v, true
}
```

The strings stored in the stack represent the state of the command. Now, we need to implement the methods of the command interface—we can start with the easiest ones:

```
func (s *Stack) GetName() string {
    return "stack"
}

func (s *Stack) GetHelp() string {
    return "a stack-like memory storage"
}
```

Now we need to decide how it works internally. There will be two sub-commands:

- push, followed by one or more arguments, will push to the stack.
- pop will take the topmost element of the stack and it will not need any argument.

Let's define a helper method, isValid, that checks whether the arguments are valid:

```
func (s *Stack) isValid(cmd string, args []string) bool {
    switch cmd {
    case "pop":
        return len(args) == 0
    case "push":
        return len(args) > 0
    default:
        return false
    }
}
```

Now, we can implement the command execution method that will use the validity check. If this is passed, it will execute the selected command or show a help message:

```
func (s *Stack) Run(r io.Reader, w io.Writer, args ...string) (exit bool) {
    if l := len(args); l < 2 || !s.isValid(args[1], args[2:]) {
        fmt.Fprintf(w, "Use `stack push <something>` or `stack pop`\n")
        return false
    }
    if args[1] == "push" {
        s.push(args[2:]...)
        return false
    }
    if v, ok := s.pop(); !ok {
```

```
        fmt.Fprintf(w, "Empty!\n")
    } else {
        fmt.Fprintf(w, "Got: `%s`\n", v)
    }
    return false
}
```

Persistent status

The next step is to persist the status between sessions, and this requires some action to be executed when the application starts and another one to be executed when the application ends. These new behaviors could be integrated with some changes on the command interface:

```
type Command interface {
    Startup() error
    Shutdown() error
    GetName() string
    GetHelp() string
    Run(r io.Reader, w io.Writer, args ...string) (exit bool)
}
```

The Startup() method is responsible for the status and loading it when the application starts, and the Shutdown() method needs to save the current status to the disk before exit. We can update the Base structure with these methods; however, this will not do anything, because there's no status:

```
// Startup does nothing
func (b Base) Startup() error { return nil }

// Shutdown does nothing
func (b Base) Shutdown() error { return nil }
```

The command list is not exported; it is the unexported variable, commands. We can add two functions that will interact with such a list, and make sure we execute these methods, Startup and Shutdown, on all available commands:

```
// Shutdown executes shutdown for all commands
func Shutdown(w io.Writer) {
    for _, c := range commands {
        if err := c.Shutdown(); err != nil {
            fmt.Fprintf(w, "%s: shutdown error: %s", c.GetName(), err)
        }
    }
}
```

```
// Startup executes Startup for all commands
func Startup(w io.Writer) {
    for _, c := range commands {
        if err := c.Startup(); err != nil {
            fmt.Fprintf(w, "%s: startup error: %s", c.GetName(), err)
        }
    }
}
```

The final step is using these functions within the main application before starting the main loop:

```
func main() {
    s, w, a, b := bufio.NewScanner(os.Stdin), os.Stdout, args{},
bytes.Buffer{}
    command.Startup(w)
    defer command.Shutdown(w) // this is executed before returning
    fmt.Fprint(w, "** Welcome to PseudoTerm! **\nPlease enter a
command.\n")
    for {
        // main loop
    }
}
```

Upgrading the Stack command

We want the command we defined before, Stack, to be able to save its state between sessions. The simplest solution is to save the contents of the stack as a text file, with one element per line. We can make this file unique for each user, using the OS/user package to place it in the user home directory:

```
func (s *Stack) getPath() (string, error) {
    u, err := user.Current()
    if err != nil {
        return "", err
    }
    return filepath.Join(u.HomeDir, ".stack"), nil
}
```

Let's start writing; we will create and truncate the file (setting its size to 0 using the TRUNC flag) and write the following lines:

```
func (s *Stack) Shutdown(w io.Writer) error {
    path, err := s.getPath()
    if err != nil {
        return err
```

```
    }
    f, err := os.OpenFile(path, os.O_CREATE|os.O_WRONLY|os.O_TRUNC, 0600)
    if err != nil {
        return err
    }
    defer f.Close()
    for _, v := range s.data {
        if _, err := fmt.Fprintln(f, v); err != nil {
            return err
        }
    }
    return nil
}
```

The method used during the shutdown will read the file line by line and will add the elements to the stack. We can use bufio.Scanner, as we saw in a previous chapter, to do this easily:

```
func (s *Stack) Startup(w io.Writer) error {
    path, err := s.getPath()
    if err != nil {
        return err
    }
    f, err := os.Open(path)
    if err != nil {
        if os.IsNotExist(err) {
            return nil
        }
        return err
    }
    defer f.Close()
    s.data = s.data[:0]
    scanner := bufio.NewScanner(f)
    for scanner.Scan() {
        s.push(string(scanner.Bytes()))
    }
    return nil
}
```

Summary

In this chapter, we went through some terminology, in order to understand why modern Terminal applications exist and how they evolved.

Then, we focused on how to implement a basic pseudo-terminal. The first step was to create a loop that handled input management, then it was necessary to create a command selector and finally an executor. The selector could choose between a series of functions defined in the package, and we created a special command to exit the application. With some refactoring, we went from functions to structs containing both the name and action.

We saw how to improve the application in various ways. First, we created a support for multiline input (using a custom split function for a scanner) that supported quoted strings, with new lines. Then, we created some tools to add colored output to our functions and used them in one of the commands previously defined. We also used Levenshtein distance to suggest similar commands when the user specifies a non-existing one.

Finally, we separated commands from the main application and created a way of registering new commands from the outside. We used an interface because this allows better extension and customization, together with a basic implementation of the interface.

In the next chapter, we will start talking about process properties and child processes.

Questions

1. What is a Terminal, and what is a pseudo-terminal?
2. What should a pseudo-terminal be able to do?
3. What Go tools did we use in order to emulate a Terminal?
4. How can my application get instructions from standard input?
5. What is the advantage of using interfaces for commands?
6. What's the Levenshtein distance? Why can it be useful in pseudo-terminals?

3

Section 3: Understanding Process Communication

This section explores how various processes communicate with each other. It explains how to use Unix's pipe-based communication in Go, how to handle signals inside an application, and how to use the network to communicate effectively. Finally, it shows how to encode data to improve communication speed.

This section consists of the following chapters:

7
Handling Processes and Daemons

This chapter will introduce you to how to handle the properties of the current process using the Go standard library, and how to change them. We will also focus on how to create child processes and give an overview of the `os/exec` package.

Finally, we will explain what daemons are, what properties they have, and how to create them using the standard library.

The following topics will be covered in this chapter:

- Understanding processes
- Child processes
- Beginning with daemons
- Creating a service

Technical requirements

This chapter requires Go to be installed and your favorite editor to be set up. For more information, you can refer to `Chapter 3`, *An Overview of Go*.

Understanding processes

We have already seen the importance of processes in the Unix operating system, so now we will look at how to obtain information on the current process and how to create and handle child processes.

Current process

The Go standard library allows us to get information on the current process. This is done by using a series of functions that are available in the `os` package.

Standard input

The first two things that a program may want to know are its identifier and the parent identifier, that is, PID and PPID. This is actually pretty straightforward – the `os.Getpid()` and `os.Getppid()` functions both return an integer value with both of the identifiers, as shown in the following code:

```go
package main

import (
    "fmt"
    "os"
)

func main() {
    fmt.Println("Current PID:", os.Getpid())
    fmt.Println("Current Parent PID:", os.Getppid())
}
```

The full example is available at `https://play.golang.org/p/ng0m9y4LcD5`.

User and group ID

Another piece of information that can be handy is the current user and the groups that the process belongs to. A typical user case could be to compare them with a file-specific permission.

The `os` package offers the following functions:

- `os.Getuid()`: Returns the user ID of the process owner
- `os.Getgid()`: Returns the group ID of the process owner
- `os.Getgroups()`: Returns additional group IDs of the process owner

We can see that these three functions return IDs in their numerical form:

```
package main

import (
    "fmt"
    "os"
)

func main() {
    fmt.Println("User ID:", os.Getuid())
    fmt.Println("Group ID:", os.Getgid())
    groups, err := os.Getgroups()
    if err != nil {
        fmt.Println(err)
        return
    }
    fmt.Println("Group IDs:", groups)
}
```

The full example is available at https://play.golang.org/p/EqmonEEc_ZI.

In order to get the names of users and groups, there are some helper functions in the os/user package. These functions (with a pretty self-explanatory name) are as follows:

- func LookupGroupId(gid string) (*Group, error)
- func LookupId(uid string) (*User, error)

Even if the user ID is an integer, it takes a string as an argument, and so a conversion needs to be done. The easiest way to do that is to use the strconv package, which offers a series of utilities to convert from strings into the other basic data types, and vice versa.

We can see them in action in the following example:

```
package main

import (
    "fmt"
    "os"
    "os/user"
    "strconv"
)

func main() {
    uid := os.Getuid()
    u, err := user.LookupId(strconv.Itoa(uid))
    if err != nil {
```

```
        fmt.Println("Error:", err)
        return
    }
    fmt.Printf("User: %s (uid %d)\n", u.Username, uid)
    gid := os.Getgid()
    group, err := user.LookupGroupId(strconv.Itoa(gid))
    if err != nil {
        fmt.Println("Error:", err)
        return
    }
    fmt.Printf("Group: %s (uid %d)\n", group.Name, uid)
}
```

The full example is available at `https://play.golang.org/p/C6EWF2c50DT`.

Working directory

Another very useful piece of information that a process can give us access to is the working directory so that we can change it. In Chapter 4, *Working with the Filesystem*, we learned about which tools we can use – `os.Getwd` and `os.Chdir`.

In the following practical example, will look at how to use these functions to manipulate the working directory:

1. First, we will obtain the current working directory and use it to get the path of the binary.
2. Then, we will join the working directory with another path and use it create a directory.
3. Finally, we will use the path of the directory we just created to change the current working directory.

Check out the following code:

```
// obtain working directory
wd, err := os.Getwd()
if err != nil {
    fmt.Println("Error:", err)
    return
}
fmt.Println("Working Directory:", wd)
fmt.Println("Application:", filepath.Join(wd, os.Args[0]))

// create a new directory
d := filepath.Join(wd, "test")
if err := os.Mkdir(d, 0755); err != nil {
```

```
        fmt.Println("Error:", err)
        return
    }
    fmt.Println("Created", d)

    // change the current directory
    if err := os.Chdir(d); err != nil {
        fmt.Println("Error:", err)
        return
    }
    fmt.Println("New Working Directory:", d)
```

The full example is available at `https://play.golang.org/p/UXAer5nGBtm`.

Child processes

A Go application can interact with the operating system to create some other processes. Another subpackage of `os` offers the functionality to create and run new processes. Inside the `os/exec` package, there is the `Cmd` type, which represents command execution:

```
type Cmd struct {
    Path string // command to run.
    Args []string // command line arguments (including command)
    Env []string // environment of the process
    Dir string // working directory
    Stdin io.Reader // standard input`
    Stdout io.Writer // standard output
    Stderr io.Writer // standard error
    ExtraFiles []*os.File // additional open files
    SysProcAttr *syscall.SysProcAttr // os specific attributes
    Process *os.Process // underlying process
    ProcessState *os.ProcessState // information on exited processte
}
```

The easiest way to create a new command is by using the `exec.Command` function, which takes the executable path and a series of arguments. Let's look at a simple example with an `echo` command and some arguments:

```
package main

import (
    "fmt"
    "os/exec"
)

func main() {
```

```
        cmd := exec.Command("echo", "A", "sample", "command")
        fmt.Println(cmd.Path, cmd.Args[1:]) // echo [A sample command]
    }
```

The full example is available at `https://play.golang.org/p/dBIAUteJbxI`.

One very important detail is the nature of standard input, output, and error – they are all interfaces that we are already familiar with:

- The input is an `io.Reader`, which could be `bytes.Reader`, `bytes.Buffer`, `strings.Reader`, `os.File`, or any other implementation.
- The output and the error are `io.Writer`, can also be `os.File` or `bytes.Buffer`, and can also be `strings.Builder` or any another writer implementation.

There are different ways to launch the process, depending on what the parent application needs:

- `Cmd.Run`: Executes the command, and returns an error that is `nil` if the child process is executed correctly.
- `Cmd.Start` : Executes the command asynchronously and lets the parent continue its flow. In order to wait for the child process to finish its execution, there is another method, `Cmd.Wait`.
- `Cmd.Output`: Executes the command and returns its standard output, and returns an error if `Stderr` isn't defined but the standard error produced the output.
- `Cmd.CombinedOutput`: Executes the command and returns both a standard error and output combined, which is very useful when the entire output of the child process-standard output plus standard error needs to be checked or saved.

Accessing child properties

Once the command has started its execution, synchronously or not, the underlying `os.Process` gets populated and it is possible to see its PID, as we can see in the following example:

```
package main

import (
    "fmt"
    "os/exec"
)
```

```
func main() {
    cmd := exec.Command("ls", "-l")
    if err := cmd.Start(); err != nil {
        fmt.Println(err)
        return
    }
    fmt.Println("Cmd: ", cmd.Args[0])
    fmt.Println("Args:", cmd.Args[1:])
    fmt.Println("PID: ", cmd.Process.Pid)
    cmd.Wait()
}
```

Standard input

Standard input can be used to send some data from the application to the child process. A buffer can be used to store the data and let the command read it, as shown in the following example:

```
package main

import (
    "bytes"
    "fmt"
    "os"
    "os/exec"
)

func main() {
    b := bytes.NewBuffer(nil)
    cmd := exec.Command("cat")
    cmd.Stdin = b
    cmd.Stdout = os.Stdout
    fmt.Fprintf(b, "Hello World! I'm using this memory address: %p", b)
    if err := cmd.Start(); err != nil {
        fmt.Println(err)
        return
    }
    cmd.Wait()
}
```

Beginning with daemons

In Unix, all of the programs that are running in the background are called **daemons**. They usually have a name that ends with the letter *d*, like `sshd` or `syslogd`, and they provide many functionalities of the OS.

Operating system support

In macOS, Unix, and Linux, a process is a daemon if it survives its parent life cycle, that is, when the parent process terminates its execution and the child process lives on. That's because the process parent is changed to the `init` process, a special daemon with no parent, and PID 1, which starts and terminates with the OS. Before going into this further, let's introduce two very important concepts – *sessions* and *process groups*:

- A process group is a collection of processes that share signal handling. The first process of the group is called the **group leader**. There is a Unix system call, `setpgid`, that is capable of changing the group for a process, but it has some limits. The process can change its own process group, or change the group of one of its child processes before the `exec` system call is executed on it. When the process group changes, the session group needs to be changed accordingly, the same as for the leader of the destination group.

- A session is a collection of process groups that allow us to impose a series of restrictions on process groups and other operations. A session doesn't allow process group migration to another session, and it prevents processes from creating process groups in different sessions. The `setsid` system call allows us to change the process session to a new session if the process isn't a process group leader. Also, the first process group ID sets the session ID. If this ID is the same as the one of the running process, that process is called the **session leader**.

Now that we've explained these two properties, we can look at the standard operations that are needed to create a daemon, which are usually the following:

- Clean up the environment to remove unnecessary variables.
- Create a fork so that the main process can terminate the process normally.
- Use the `setsid` system call, which accomplishes these three steps:
    1. Remove the PPID from the forked process so that it gets adopted by the `init` process
    2. Create a new session for the fork, which will become the session leader
    3. Set the process as the group leader

- The fork's current directory is set to the root directory to avoid having other directories in use, and all the files opened by the parent are closed (the child will open them if needed).
- Setting standard input to `/dev/null` and using some log files as standard output and error.
- Optionally, the fork can be forked again, and can then exit. The first fork will be the group leader, and the second one will have the same group, allowing us to have another fork that isn't a group leader.

This is valid for Unix-based operating systems, though Windows also has support for permanent background processes, which are called **services**. Services can start automatically at boot or be started and stopped manually using a visual application called **Service Control Manager (SCM)**. They can also be controlled from a command line, using the `sc` command in the regular prompt, and via the `Start-Service` and `Stop-Service` cmdlets in PowerShell.

Daemons in action

Now that we understand what a daemon is and how it works, we can attempt to use the Go standard library to create one. Go applications are multithreaded and don't allow us to call to the `fork` system call directly.

We have learned that the `Cmd.Start` method in the `os/exec` package allows us to start a process asynchronously. The second step is to use the `release` method to close all the resources from the current process.

The following example shows us how to do this:

```
package main

import (
    "fmt"
    "os"
    "os/exec"
    "time"
)

var pid = os.Getpid()

func main() {
    fmt.Printf("[%d] Start\n", pid)
    fmt.Printf("[%d] PPID: %d\n", pid, os.Getppid())
    defer fmt.Printf("[%d] Exit\n\n", pid)
```

```
    if len(os.Args) != 1 {
        runDaemon()
        return
    }
    if err := forkProcess(); err != nil {
        fmt.Printf("[%d] Fork error: %s\n", pid, err)
        return
    }
    if err := releaseResources(); err != nil {
        fmt.Printf("[%d] Release error: %s\n", pid, err)
        return
    }
}
```

Let's see what the `forkProcess` function does, create another process, and start it:

1. First, the process working directory gets set to root, and the output and error streams are set to the standard ones:

```
func forkProcess() error {
    cmd := exec.Command(os.Args[0], "daemon")
    cmd.Stdout, cmd.Stderr, cmd.Dir = os.Stdout, os.Stderr, "/"
    return cmd.Start()
}
```

2. Then, we can release the resources – first, though, we need to find the current process. Then, we can call the `os.Process` method, `Release`, to ensure that the main process releases its resources:

```
func releaseResources() error {
    p, err := os.FindProcess(pid)
    if err != nil {
        return err
    }
    return p.Release()
}
```

3. The `main` function will contain the daemon logic, which in this example is very simple – it will just print what is running every few seconds:

```
func runDaemon() {
    for {
        fmt.Printf("[%d] Daemon mode\n", pid)
        time.Sleep(time.Second * 10)
    }
}
```

Services

We have already seen how the first process that lives from boot to the shutdown of the OS is called `init` or `init.d` since it's a daemon. This process is responsible for handling the other daemons, and stores its configuration in the `/etc/init.d` directory.

Each Linux distribution uses its own version of the daemon control process, like `upstart` for Chrome OS or `systemd` in Arch Linux. They all serve the same purpose and have similar behavior.

Each daemon has a control script or application that resides inside `/etc/init.d` and should be able to interpret a series of commands as first arguments, such as `status`, `start`, `stop`, and `restart`. In most cases, the `init.d` file is a script that executes a switch on the argument and behaves accordingly.

Creating a service

Some applications are capable of automatically handling their service file, and this is what we will try to achieve, step by step. Let's start with an `init.d` script:

```
#!/bin/sh

"/path/to/mydaemon" $1
```

This is a sample script that passes the first argument to the daemon. The path to the binary will be dependent on where the file is located. This needs to be defined at runtime:

```go
// ErrSudo is an error that suggest to execute the command as super user
// It will be used with the functions that fail because of permissions
var ErrSudo error

var (
    bin string
    cmd string
)

func init() {
    p, err := filepath.Abs(filepath.Dir(os.Args[0]))
    if err != nil {
        panic(err)
    }
    bin = p
    if len(os.Args) != 1 {
```

```
            cmd = os.Args[1]
        }
        ErrSudo = fmt.Errorf("try `sudo %s %s`", bin, cmd)
    }
```

The `main` function will handle the different commands, as follows:

```
func main() {
    var err error
    switch cmd {
    case "run":
        err = runApp()
    case "install":
        err = installApp()
    case "uninstall":
        err = uninstallApp()
    case "status":
        err = statusApp()
    case "start":
        err = startApp()
    case "stop":
        err = stopApp()
    default:
        helpApp()
    }
    if err != nil {
        fmt.Println(cmd, "error:", err)
    }
}
```

How can we make sure that our app is running? A very sound strategy is to use a `PID` file, which is a text file that contains the current PID of the running process. Let's define a couple of auxiliary functions to achieve this:

```
const (
    varDir = "/var/mydaemon/"
    pidFile = "mydaemon.pid"
)

func writePid(pid int) (err error) {
    f, err := os.OpenFile(filepath.Join(varDir, pidFile),
os.O_CREATE|os.O_WRONLY, 0644)
    if err != nil {
        return err
    }
    defer f.Close()
    if _, err = fmt.Fprintf(f, "%d", pid); err != nil {
        return err
```

```
    }
    return nil
}

func getPid() (pid int, err error) {
    b, err := ioutil.ReadFile(filepath.Join(varDir, pidFile))
    if err != nil {
        return 0, err
    }
    if pid, err = strconv.Atoi(string(b)); err != nil {
        return 0, fmt.Errorf("Invalid PID value: %s", string(b))
    }
    return pid, nil
}
```

The `install` and `uninstall` functions will take care of adding or removing the service file located at `/etc/init.d/mydaemon` and requires us to launch the app with root permissions because of the file's location:

```
const initdFile = "/etc/init.d/mydaemon"

func installApp() error {
    _, err := os.Stat(initdFile)
    if err == nil {
        return errors.New("Already installed")
    }
    f, err := os.OpenFile(initdFile, os.O_CREATE|os.O_WRONLY, 0755)
    if err != nil {
        if !os.IsPermission(err) {
            return err
        }
        return ErrSudo
    }
    defer f.Close()
    if _, err = fmt.Fprintf(f, "#!/bin/sh\n\"%s\" $1", bin); err != nil {
        return err
    }
    fmt.Println("Daemon", bin, "installed")
    return nil
}

func uninstallApp() error {
    _, err := os.Stat(initdFile)
    if err != nil && os.IsNotExist(err) {
        return errors.New("not installed")
    }
    if err = os.Remove(initdFile); err != nil {
        if err != nil {
```

```
                    if !os.IsPermission(err) {
                        return err
                    }
                return ErrSudo
                }
            }
        fmt.Println("Daemon", bin, "removed")
        return err
    }
```

Once the file is created, we can install the app as a service with the `mydaemon` `install` command and remove it with `mydaemon uninstall`.

Once the daemon has been installed, we can use `sudo service mydaemon` `[start|stop|status]` to control the daemon. Now, all we need to do is implement these actions:

- `status` will look for the `pid` file, read it, and send a signal to the process to check whether it's running.
- `start` will run the application with the `run` command and write the `pid` file.
- `stop` will get the `pid` file, find the process, kill it, and then remove the `pid` file.

Let's take a look at how the `status` command is implemented. Note that the `0` signal doesn't exist in Unix, and doesn't trigger an action from the operating system or the app, but the operation will fail if the process isn't running. This tells us whether the process is alive or not:

```
func statusApp() (err error) {
    var pid int
    defer func() {
        if pid == 0 {
            fmt.Println("status: not active")
            return
        }
        fmt.Println("status: active - pid", pid)
    }()
    pid, err = getPid()
    if err != nil {
        if os.IsNotExist(err) {
            return nil
        }
        return err
    }
    p, err := os.FindProcess(pid)
    if err != nil {
        return nil
```

```
    }
    if err = p.Signal(syscall.Signal(0)); err != nil {
        fmt.Println(pid, "not found - removing PID file...")
        os.Remove(filepath.Join(varDir, pidFile))
        pid = 0
    }
    return nil
}
```

In the `start` command, we will create the daemon by following the steps we covered in the *Operating system support section*:

1. Use files for standard output and input
2. Set the working directory to root
3. Start the command asynchronously

In addition to these operations, the `start` command saves the PID value of the process in a specific file, which will be used to see whether the process is alive:

```
func startApp() (err error) {
    const perm = os.O_CREATE | os.O_APPEND | os.O_WRONLY
    if err = os.MkdirAll(varDir, 0755); err != nil {
        if !os.IsPermission(err) {
            return err
        }
        return ErrSudo
    }
    cmd := exec.Command(bin, "run")
    cmd.Stdout, err = os.OpenFile(filepath.Join(varDir, outFile),
        perm, 0644)
            if err != nil {
                return err
            }
    cmd.Stderr, err = os.OpenFile(filepath.Join(varDir, errFile),
        perm, 0644)
            if err != nil {
                return err
            }
    cmd.Dir = "/"
    if err = cmd.Start(); err != nil {
        return err
    }
    if err := writePid(cmd.Process.Pid); err != nil {
        if err := cmd.Process.Kill(); err != nil {
            fmt.Println("Cannot kill process", cmd.Process.Pid, err)
        }
        return err
```

```
        }
        fmt.Println("Started with PID", cmd.Process.Pid)
        return nil
    }
```

Lastly, `stopApp` will terminate the process identified by the PID file, if it exists:

```
func stopApp() (err error) {
    pid, err := getPid()
    if err != nil {
        if os.IsNotExist(err) {
            return nil
        }
        return err
    }
    p, err := os.FindProcess(pid)
    if err != nil {
        return nil
    }
    if err = p.Signal(os.Kill); err != nil {
        return err
    }
    if err := os.Remove(filepath.Join(varDir, pidFile)); err != nil {
        return err
    }
    fmt.Println("Stopped PID", pid)
    return nil
}
```

Now, all of the parts that are needed for an application's control are there, and all that is missing is the main application part, which should be a loop so that the daemon stays alive:

```
func runApp() error {
    fmt.Println("RUN")
    for {
        time.Sleep(time.Second)
    }
    return nil
}
```

In this example, all it does is sleep for a fixed amount of time, between the loop iterations. This is generally a good idea in a main loop, because an empty `for` loop would use a lot of resources for no reason. Let's assume that your application is checking for a certain condition in a `for` loop. If that is satisfied, continually checking for this will use a lot of resources. Adding an idle sleep of a few milliseconds can help reduce idle CPU consumption by 90-95%, so keep it in mind when designing your daemons!

Third-party packages

So far, we have seen how to implement daemons from scratch using the init.d service. Our implementation is very simple and limited. It could be improved, but there are already many packages that offer the same functionalities. They offer support for different providers such as init.d and systemd, and some of them also work across non-Unix OSes such as Windows.

One of the more famous packages (more than 1,000 stars on GitHub) is kardianos/service, which supports all major platforms – Linux, macOS, and Windows.

It defines a main interface that represents a daemon and has two methods – one for starting the daemon, and the other for stopping it. Both are non-blocking:

```
type Interface interface {
    // Start provides a place to initiate the service. The service doesn't not
    // signal a completed start until after this function returns, so the
    // Start function must not take more than a few seconds at most.
    Start(s Service) error

    // Stop provides a place to clean up program execution before it is terminated.
    // It should not take more than a few seconds to execute.
    // Stop should not call os.Exit directly in the function.
    Stop(s Service) error
}
```

The package already offers some use cases, from simple to more convoluted ones, in the example (https://github.com/kardianos/service/tree/master/example) directory. The best practice is to start a goroutine with the main activity loop. The Start method can be used to open and prepare the necessary resources, while Stop should be used to free them, and for other deferred activities such as buffer flushing.

Some other packages only offer compatibility with Unix systems, such as takama/daemon (https://github.com/takama/daemon), which works in a similar fashion. It also offers some usage examples.

Summary

In this chapter, we reviewed how to get information related to the current process, PID and PPID, UID and GID, and the working directory. Then, we saw how the os/exec package allows us to create child processes, and how their properties can be read similar to the current process.

Next, we looked at what daemons are and how various operating systems support them. We verified how simple it is to execute a process that outlives its parent with os/exec using Cmd.Run.

Then, we went through the automated daemon management systems that are available with Unix, and created an application capable of running with service step by step.

In the next chapter, we are going to improve the control we have over child processes by checking out how to use exit codes and how to manage and send signals.

Questions

1. What information is available for the current process inside a Go application?
2. How do you create a child process?
3. How do you ensure that a child process survives its parent?
4. Can you access child properties? How can you use them?
5. What's a daemon in Linux and how are they handled?

Exit Codes, Signals, and Pipes

8

This chapter will continue on from the previous one and will demonstrate the communication between parent and child processes. In particular, the chapter will show you how communication can be managed with the correct usage of exit codes, through custom signal handling and by connecting processes with pipes. These forms of communication will be used to allow our application to communicate effectively with the operating system and other processes.

The following topics will be covered in this chapter:

- Returning exit codes
- Reading exit codes
- Intercepting signals
- Sending signals
- Using pipes
- Using other stream utilities

Technical requirements

This chapter requires Go to be installed and your favorite editor to be set up. For more information, you can refer to Chapter 3, *An Overview of Go*.

Using exit codes

An exit code, or exit status, is a small integer number that is passed from a process to its parent when the process exits. It is the simplest way of notifying you of the result of the application execution. In Chapter 2, *Unix OS Components*, we briefly mentioned exit codes. We will now learn how to use them inside applications and how to interpret the exit codes of child processes.

Sending exit codes

Exit codes are the way in which a process notifies its parent about its status after terminating. In order to return any exit status from the current process, there is one function that does the job in a straightforward way: `os.Exit`.

This function takes one argument, that is, an integer, and represents the exit code that will be returned to the parent process. This can be verified using a simple program, as shown in the following code:

```
package main

import (
    "fmt"
    "os"
)

func main() {
    fmt.Println("Hello, playground")
    os.Exit(1)
}
```

The full example is available at `https://play.golang.org/p/-6GIY7EaVD_V`.

The exit code, `0`, is used when the application is executed successfully. Any other exit code symbolizes some type of error that may have occurred during the execution. When the main function finishes, it returns a `0`; when a panic is not recovered, it returns a `2`.

Exit codes in bash

Every time a command gets executed in a shell, the resulting exit code gets stored in a variable. The status of the last command executed gets stored in the `$?` variable, and it can be printed as follows:

```
> echo  $? # will print 1
```

It is important to note that the exit code only works when you run a binary obtained with `go build` or `go install`. If you use `go run`, it will return 1 for any code that is not `0`.

The exit value bit size

The exit status is an 8-bit integer; this means that even if the argument for the Go function is an integer, then the status that is returned will be the result of a modulo operation between the passed value and 256.

Let's take a look at the following program:

```
package main

import (
    "fmt"
    "os"
)

func main() {
    fmt.Println("Hello, playground")
    os.Exit(-1)
}
```

The full example is available at https://play.golang.org/p/vzwI1kDiGrP.

This will have an exit status of 255 even if the function argument is −1 because (−1)%256=255. This happens because the exit code is an 8-bit number (0, 255).

Exit and deferred functions

One important caveat about this function's use is that deferred functions are not executed.

The following example will have no output:

```
package main

import (
    "fmt"
    "os"
)

func main() {
    defer fmt.Println("Hello, playground")
    os.Exit(0)
}
```

The full example is available at https://play.golang.org/p/2zbczc_ckgb.

Panics and exit codes

If the application terminates for a panic that is not recovered, then the deferred function will be executed, but the exit code will be 2:

```
package main

import (
    "fmt"
)

func main() {
    defer fmt.Println("Hello, playground")
    panic("panic")
}
```

The full example is available at https://play.golang.org/p/mjOMb0KsM3e.

Exit codes and goroutines

If the os.Exit function happens in a goroutine, all the goroutines (including the main one) will terminate immediately without executing any deferred call, as follows:

```
package main

import (
    "fmt"
    "os"
    "time"
)

func main() {
    go func() {
        defer fmt.Println("go end (deferred)")
        fmt.Println("go start")
        os.Exit(1)
    }()
    fmt.Println("main end (deferred)")
    fmt.Println("main start")
    time.Sleep(time.Second)
    fmt.Println("main end")
}
```

The full example is available at `https://play.golang.org/p/JVEB5MTcEoa`.

It is necessary for you to use `os.Exit` carefully because, since all the deferred operations will not be executed, this could lead to resource leaks or errors, such as not flushing a buffer and not writing all the content to a file.

Reading child process exit codes

We explored how to create child processes in the previous chapter. Go makes it possible for you to easily check child exit codes, however, it's not straightforward because there is a field in the `exec.Cmd` struct that has an `os.ProcessState` attribute.

The `os.ProcessState` attribute has a `Sys` method that returns an interface. Its value is a `syscall.WaitStatus` struct in Unix, which makes it possible to access the exit code using the `ExitCode` method. This is demonstrated using the following code:

```
package main

import (
    "fmt"
    "os"
    "os/exec"
    "syscall"
)

func exitStatus(state *os.ProcessState) int {
    status, ok := state.Sys().(syscall.WaitStatus)
    if !ok {
        return -1
    }
    return status.ExitStatus()
}

func main() {
    cmd := exec.Command("ls", "__a__")
    if err := cmd.Run(); err != nil {
        if status := exitStatus(cmd.ProcessState); status == -1 {
            fmt.Println(err)
        } else {
            fmt.Println("Status:", status)
        }
    }
}
```

If the command variable cannot be accessed, then the error returned is `exec.ExitError` and this wraps the `os.ProcessState` attribute, as follows:

```
func processState(e error) *os.ProcessState {
    err, ok := e.(*exec.ExitError)
    if !ok {
        return nil
    }
    return err.ProcessState
}
```

We can see that obtaining the exit code is not straightforward and requires some typecasting.

Handling signals

Signals are the other inter-process communication tools that Unix operating systems offer. They are integer values that can be sent from one process to another process, giving our application the capability of communicating with more processes than just the parent. By doing so, the application is capable of interpreting the incoming signals and can also send signals to other processes.

Handling incoming signals

The normal behavior for a Go application is to handle some incoming signals, including `SIGHUP`, `SIGINT`, and `SIGABRT`, by terminating the application. We can replace this standard behavior with a custom behavior that intercepts all or some of the signals and acts accordingly.

The signal package

The custom behavior is accomplished using the `os/signal` package, which exposes the necessary functions.

For instance, if the application does not need to intercept a signal, the `signal.Ignore` function allows you to add a signal to the ignored list. The `signal.Ignored` function also permits you to verify whether a certain signal has been ignored.

In order to intercept signals with a channel, you can use the core function, that is, `signal.Notify`. This makes it possible to specify a channel and to choose which signals should be sent to that channel. The application can then use the channel in any goroutine to handle signals with a custom behavior. Note that if no signal is specified, then the channel will receive all the signals that are sent to the application, as follows:

```
signal.Notify(ch, signalList...)
```

The `signal.Stop` function is used to halt the reception of signals from a certain channel, while the `signal.Reset` function stops intercepting one or more signals to all channels. In order to reset all the signals, no argument needs to be passed to `Reset`.

Graceful shutdowns

An application executes a graceful shutdown when it waits for a task to finish and clears all of its resources before terminating. Using custom signal handling is a very good practice, because it gives us time to release resources that are still open. Before shutting down, we can perform any other task that should be done before exiting the application; for instance, saving the current status.

Now that we know how exit codes work, we can introduce the `log` package. This will be used from now on to print a statement to the standard output instead of `fmt`. This makes it possible to execute the `Print` statement and the `Fatal` statement, which is the equivalent of printing and executing `os.Exit(1)`. The `log` package also allows the user to define log flags to print the date, time, and/or file/line with each print.

We can start with a very basic example, handling all signals as follows:

```
package main

import (
    "log"
    "os"
    "os/signal"
    "syscall"
)
```

```go
func main() {
    log.Println("Start application...")
    c := make(chan os.Signal)
    signal.Notify(c)
    s := <-c
    log.Println("Exit with signal:", s)
}
```

In order to test this application, you can use two different Terminals. First, you can launch the application in the first Terminal and use the other Terminal to execute the ps command to find the PID application, in order to send signals to it with the kill command.

The second method, which uses just one Terminal, launches the application in the background. This will show the PID on the screen and will be used in the kill command, as follows:

```
$ go build -o "signal" ch8/signal/base/base.go

$ ./signal &
[1] 265
[Log] Start application...

$ kill -6 265
[Log] Exit with signal: aborted
```

Note that if you're using macOS, you'll get an abort trap signal name.

Exit cleanup and resource release

A more practical and common example of a clean shutdown is resource cleanup. When using exit statements, deferred functions such as Flush for a bufio.Writer struct are not executed. This can lead to information loss, as shown in the following example:

```go
package main

import (
    "bufio"
    "fmt"
    "log"
    "os"
    "time"
)

func main() {
    f, err := os.OpenFile("file.txt", os.O_CREATE|os.O_TRUNC|os.O_WRONLY,
0644)
```

```
    if err != nil {
        log.Fatal(err)
    }
    defer f.Close()
    w := bufio.NewWriter(f)
    defer w.Flush()
    for i := 0; i < 3; i++ {
        fmt.Fprintln(w, "hello")
        log.Println(i)
        time.Sleep(time.Second)
    }
}
```

If a TERM signal is sent to this application before it finishes, the file will be created and truncated, but the flush will never be executed—resulting in an empty file.

This could, perhaps, be the intended behavior, but this is rarely the case. It is better to do any cleanup in the signal handling part, as shown in the following example:

```
func main() {
    c := make(chan os.Signal, syscall.SIGTERM)
    signal.Notify(c)
    f, err := os.OpenFile("file.txt", os.O_CREATE|os.O_TRUNC|os.O_WRONLY,
0644)
    if err != nil {
        log.Fatal(err)
    }
    defer f.Close()
    w := bufio.NewWriter(f)
    go func() {
        <-c
        w.Flush()
        os.Exit(0)
    }()
    for i := 0; i < 3; i++ {
        fmt.Fprintln(w, "hello")
        log.Println(i)
        time.Sleep(time.Second)
    }
}
```

In this case, we are using a goroutine in combination with the signal channel to flush the writer before exiting. This will ensure that whatever is written to the buffer gets persisted on the file.

Configuration reload

Signals can be used for more than just terminating an application. An application can react differently to each signal, so that it can be used to accomplish different functions, making it possible to control the application flow.

The next example will have some settings stored in a text file. The settings will be of the time.Duration type stored as its string version. A duration is an int64 value with a string version that is in a human-readable format, such as 2m10s, which also has many useful methods. This is used in different functions of the time package.

The app will execute a certain operation with a frequency that depends on the current settings value. Possible operations with signals include the following:

- SIGHUP (1): This loads the interval from the settings file.
- SIGTERM (2): This saves the current interval values, and exits the application.
- SIGQUIT (6): This exits without saving.
- SIGUSR1 (10): This increases the interval by doubling it.
- SIGUSR2 (11): This decreases the interval by halving it.
- SIGALRM (14): This saves the current interval value.

Capturing these signals is done with the signal.Notify function, which is called for all the different signals. The value received from the channel requires a condition statement, a type switch, to allow the application to execute a different action according to the value:

```go
func main() {
    c := make(chan os.Signal, 1)
    d := time.Second * 4
    signal.Notify(c,
        syscall.SIGHUP, syscall.SIGINT, syscall.SIGQUIT,
        syscall.SIGUSR1, syscall.SIGUSR2, syscall.SIGALRM)
    // initial load
    if err := handleSignal(syscall.SIGHUP, &d); err != nil &&
        !os.IsNotExist(err) {
            log.Fatal(err)
    }

    for {
        select {
        case s := <-c:
            if err := handleSignal(s, &d); err != nil {
                log.Printf("Error handling %s: %s", s, err)
                continue
            }
```

```
            default:
                time.Sleep(d)
                log.Println("After", d, "Executing action!")
            }
        }
    }
```

The handleSignal function will contain the switch statement in the signal:

```
func handleSignal(s os.Signal, d *time.Duration) error {
    switch s {
    case syscall.SIGHUP:
        return loadSettings(d)
    case syscall.SIGALRM:
        return saveSettings(d)
    case syscall.SIGINT:
        if err := saveSettings(d); err != nil {
            log.Println("Cannot save:", err)
            os.Exit(1)
        }
        fallthrough
    case syscall.SIGQUIT:
        os.Exit(0)
    case syscall.SIGUSR1:
        changeSettings(d, (*d)*2)
        return nil
    case syscall.SIGUSR2:
        changeSettings(d, (*d)/2)
        return nil
    }
    return nil
}
```

The following describes the different behaviors that will be implemented in the signal handing function:

- Changing the value will just use the duration pointer to store the new value.
- Loading will try to scan the content of the file (if present) as a duration and change the settings value.
- Saving will write the duration to the file, and it will use its string format. The following code depicts this:

```
func changeSettings(d *time.Duration, v time.Duration) {
    *d = v
    log.Println("Changed", v)
}
```

```go
func loadSettings(d *time.Duration) error {
    b, err := ioutil.ReadFile(cfgPath)
    if err != nil {
        return err
    }
    var v time.Duration
    if v, err = time.ParseDuration(string(b)); err != nil {
        return err
    }
    *d = v
    log.Println("Loaded", v)
    return nil
}

func saveSettings(d *time.Duration) error {
    f, err := os.OpenFile(cfgPath,
        os.O_WRONLY|os.O_CREATE|os.O_TRUNC, 0644)
            if err != nil {
                return err
            }
        defer f.Close()

    if _, err = fmt.Fprint(f, d); err != nil {
        return err
    }
    log.Println("Saved", *d)
    return nil
```

We will get the path of the user's home directory in the `init` function and use it to compose the path of the `settings` file, as follows:

```go
var cfgPath string

func init() {
    u, err := user.Current()
    if err != nil {
        log.Fatalln("user:", err)
    }
    cfgPath = filepath.Join(u.HomeDir, ".multi")
}
```

We can launch the application in a Terminal and use another Terminal to send signals, as follows:

Terminal 1	Terminal 2

`$ go run ch08/signal/multi/multi.go`
`Loaded 1s`
`After 1s Executing action!`	

`Changed 2s` `After 2s Executing action!`	`$ kill -SIGUSR1 $(pgrep multi)`
`Changed 4s` `After 4s Executing action!`	`$ kill -SIGUSR1 $(pgrep multi)`
`Changed 2s` `After 2s Executing action!`	`$ kill -SIGUSR2 $(pgrep multi)`

`Saved 1s`
 	
`$`
	`$ kill -SIGINT $(pgrep multi)`

In the left column, we can see the output from the application; in the right column, we can see the commands we launched to trigger them. In order to get the PID of the running application, we are using the `pgrep` command and nesting in `kill`.

Sending signals to other processes

After seeing the way in which incoming signals are handled, let's take a look at how to send signals to other processes programmatically. The `os.Process` structure is the only tool we need—its `Signal` method makes it possible to send a signal to the project. It's that easy!

The less simple part is obtaining the process. There are two use cases, as follows:

- The process is a child, and we have the process value already as a result of `os.StartProcess` or `exec.Command` structs.
- The process is already existing but we do not have it, so we need to search for it using its PID.

The first use case is simpler, as we already have the process as a variable, or as an attribute of an `exec.Cmd` variable, and we can call the method directly.

The other use case requires searching for the process by PID using the `os.FindProcess` method, as follows:

```
p, err := os.FindProcess(pid)
if err != nil {
    panic(err)
}
```

Once we have `os.Process`, we can use its `Signal` method to send a specific signal to it, as follows:

```
if err = p.Signal(syscall.SIGTERM); err != nil {
    panic(err)
}
```

The type of signal we will send to a process depends on the destination process and the behavior we want to suggest, such as an interruption or a termination.

Connecting streams

Streams in Go are an abstraction that makes it possible to see any sort of communication or data flow as a series of readers and writers. We have already learned that streams are an important element of Go. We'll now learn how to use what we already know about input and output to control the streams related to a process—the input, output, and error.

Pipes

Pipes are one of the best ways of connecting input and output in a synchronous way, allowing processes to communicate.

Anonymous pipes

Different commands can be chained in a sequence when using a shell, making one command output the following command input. For example, consider the following command:

```
cat book_list.txt | grep "Game" | wc -l
```

Here, we are displaying a file, using the preceding command to filter rows containing a certain string and, finally, using the filtered output to count the lines.

This can be done programmatically in Go when the processes are created inside the application.

The `io.Pipe` function returns a reader/writer couple that is connected; whatever gets written to the pipe writer will be read by the pipe reader. The write operation is a blocking one, which means that all of the written data has to be read before a new write operation can be executed.

We already saw that `exec.Cmd` allows generic streams for its output and input—this allows us to use the values returned by the `io.Pipe` function to connect one process to the other.

First, we define the three commands, as follows:

- `cat` with index 0
- `grep` with index 1
- `wc` with index 2

Then, we can define the two pipes we need, as shown in the following code:

```
r1, w1 := io.Pipe()
r2, w2 := io.Pipe()

var cmds = []*exec.Cmd{
    exec.Command("cat", "book_list.txt"),
    exec.Command("grep", "Game"),
    exec.Command("wc", "-l"),
}
```

Next, we link the input and output streams. We connect the `cat` (command 0) output and the `grep` (command 1) input, and we do the same for the `grep` output and the `wc` input:

```
cmds[1].Stdin, cmds[0].Stdout = r1, w1
cmds[2].Stdin, cmds[1].Stdout = r2, w2
cmds[2].Stdout = os.Stdout
```

We then start our commands, as follows:

```
for i := range cmds {
    if err := cmds[i].Start(); err != nil {
        log.Fatalln("Start", i, err)
    }
}
```

We wait until the end of the execution of each command and we close the respective pipe writer; otherwise, the next command reader will hang. To make things easier, each pipe writer is an element in a slice, and each writer has the same index as the command that it is linked to. The last one is `nil` because the last command is not linked by a pipe:

```
for i, closer := range []io.Closer{w1, w2, nil} {
    if err := cmds[i].Wait(); err != nil {
        log.Fatalln("Wait", i, err)
    }
    if closer == nil {
        continue
    }
    if err := closer.Close(); err != nil {
        log.Fatalln("Close", i, err)
    }
}
```

There are also other tools provided by the `io` package that can help simplify some operations.

Standard input and output pipes

The `io.MultiWriter` function makes it possible to write the same content to several readers. This will be extremely useful when the output of a command needs to be broadcast to a series of different commands automatically.

Let's say that we want to do what we did before (that is, look for a word in a file), but for different words. We can use the `MultiWriter` function to copy the output to a series of `grep` commands, each of which will be connected to its own `wc` command.

We will use two auxiliary methods of `exec.Command` in this example:

- `Cmd.StdinPipe`: This returns a `PipeWriter` struct that will be connected to the command standard input.
- `Cmd.StdoutPipe`: This returns a `PipeReader` struct that will be connected to the command standard output.

Let's start by defining a list of search terms: one for the command's tuples (`grep` and `wc`), one for writers connected to the first command, and one for the final output of each command chain:

```
var (
    words = []string{"Game", "Feast", "Dragons", "of"}
    cmds = make([][2]*exec.Cmd, len(words))
```

```
    writers = make([]io.Writer, len(words))
    buffers = make([]bytes.Buffer, len(words))
    err error
)
```

Now let's define the commands and their connection—each `grep` command will be connected on one side to `cat` using the `MultiWriter` function with the piped writers, and on the other side to the input of the `wc` command:

```
for i := range words {
    cmds[i][0] = exec.Command("grep", words[i])
    if writers[i], err = cmds[i][0].StdinPipe(); err != nil {
        log.Fatal("in pipe", i, err)
    }
    cmds[i][1] = exec.Command("wc", "-l")
    if cmds[i][1].Stdin, err = cmds[i][0].StdoutPipe(); err != nil {
        log.Fatal("in pipe", i, err)
    }
    cmds[i][1].Stdout = &buffers[i]
}

cat := exec.Command("cat", "book_list.txt")
cat.Stdout = io.MultiWriter(writers...)
```

We can run the main `cat` command and, when it finishes, we can close the first set of writing pipes, so that the `grep` command can terminate, as follows:

```
for i := range cmds {
    if err := writers[i].(io.Closer).Close(); err != nil {
        log.Fatalln("close 0", i, err)
    }
}

for i := range cmds {
    if err := cmds[i][0].Wait(); err != nil {
        log.Fatalln("grep wait", i, err)
    }
}
```

Then we can wait for the other command to finish and show the results, as follows:

```
for i := range cmds {
    if err := cmds[i][1].Wait(); err != nil {
        log.Fatalln("wc wait", i, err)
    }
    count := bytes.TrimSpace(buffers[i].Bytes())
    log.Printf("%10q %s entries", cmds[i][0].Args[1], count)
}
```

Note that, when the StdinPipe method is used, the resulting writer must be closed, but this is not necessary with StdoutPipe method.

Summary

In this chapter, we learned how to handle communications between processes using three main functionalities: exit codes, signals, and pipes.

Exit codes are 8-bit values between 0 and 255 that are returned by a process to their parent. An exit code of 0 means that the application execution was successful. It's pretty easy to return an exit code in Go, but doing so using the os.Exit function shortcuts the execution of the deferred functions. When a panic occurs, all the deferred functions are executed and the returned code is 2. Getting an exit code from a child process is relatively trickier, because it's OS-dependent; however, in Unix systems, it can be accomplished using a series of type assertions.

Signals are used to communicate with any process. They are 6-bit values, between 1 and 64, sent from one process to another using a system call. Receiving signals can be done using a channel and the signal.Notify function. Sending signals is easily done with the Process.Signal method.

Pipes are a tuple of input and output streams that are synchronously connected. They are used to connect a process input to another process output. We saw how to connect multiple commands in the same way that a Terminal does and we learned how to use io.MultiReader to broadcast one command output to many commands.

In the next chapter, we are going to dig into network programming, starting from TCP all the way up to HTTP servers.

Questions

1. What is an exit code? Who makes use of it?
2. What happens when an application panics? Which exit code is returned?
3. What is the default behavior of a Go application when receiving all signals?
4. How do you intercept signals and decide how the application must behave?
5. Can you send signals to other processes? If so, how?
6. What are pipes and why are they important?

Network Programming

9

This chapter will cover network programming. This will allow our applications to communicate with other programs that are running on any remote computer, on the same local network, or even on the internet.

We will start with some theory about network and architectures. We will then discuss socket-level communication and explain how to create a web server. Finally, we will discuss how the Go built-in template engine works.

The following topics will be covered in this chapter:

- Networking
- Socket programming
- Web servers
- Template engine

Technical requirements

This chapter requires Go to be installed and your favorite editor to be set up. For more information, you can refer to Chapter 3, *An Overview of Go*.

Additionally, it requires OpenSSL to be installed on your machine. Many Linux distributions are already shipped with some OpenSSL versions. It can also be installed on Windows, using the official installer or a third-party package manager, such as Chocolatey or Scoop.

Communicating via networks

Communication between applications can happen through a network, even if the applications are on the same machine. In order to transfer information, they need to establish a common protocol that specifies what happens all the way from the application to the wire.

OSI model

The **Open Systems Interconnect (OSI)** model is a theoretical model that dates back to the early 1970s. It defines a standard of communication that works regardless of the physical or technical structure of a network, with the goal of providing interoperability for different networks.

The model defines seven different layers, numbered from one to seven, and each layer has a higher level of abstraction to the previous one. The first three layers are often referred to as **media layers**, while the other four are the **host layers**. Let's examine each, one by one, in the following sections.

Layer 1 – Physical layer

The very first layer of the OSI model is the physical one, which is in charge of the transmission of unprocessed data from a device, similar to an Ethernet port, and its transmission medium, such as an Ethernet cable. This layer defines all the characteristics that are related to the physical/material nature of the connection—size, shape, voltages of the connector, frequencies, and timings.

Another aspect defined by the physical layer is the direction of the transmission, which can be one of the following:

- **Simplex**: The communication is one way.
- **Half duplex**: The communication is two way, but the communication happens only in one direction at a time.
- **Full duplex**: Two way-communication, where both ends can communicate at the same time.

Many well-known technologies, including Bluetooth and Ethernet, include a definition of the physical layer they are using.

Layer 2 – Data link layer

The next layer is the data link, which defines how the data transfer should happen between two nodes that are directly connected. It is responsible for the following:

- Detection of communication errors in the first layer
- Correction of the physical errors
- Control of the flow/transmission rate between nodes
- Connection termination

Some real-world examples of data link layers definition are the ones for Ethernet (802.3) and Wi-Fi (802.11).

Layer 3 – Network layer

The network layer is the next layer, and it focuses on sequences of data, called packets, that can have a variable length. A packet is transmitted from one node to another, and these two nodes could be located on the same network or on different ones.

This layer defines the network as a series of nodes connected to the same medium, identified by the previous two layers. The network is capable of delivering a message, knowing only its destination address.

Layer 4 – Transport layer

The fourth layer is transport, which ensures that packets go from sender to receiver. This is achieved with **acknowledgement (ACK)** and **negative-acknowledgement (NACK)** messages from the destination, which can trigger a repetition of the messages, until they are received correctly. There are also other mechanisms in play, such as splitting a message into chunks for transmission (segmentation), reassembling the parts in a single message (desegmentation), and checking if the data was sent and received successfully (error control).

The OSI model specifies five different transport protocols—TP0, TP1, TP2, TP3, and TP4. TP0 is the simplest one, which executes only the segmentation and reassembly of messages. The other classes add other functionalities on top of it—for instance, retransmission or timeouts.

Layer 5 – Session layer

The fifth layer introduces the concept of sessions, which is the temporary interactive exchange of information between two computers. It is responsible for creating connections and terminating them (while keeping track of the sessions) and allowing checkpointing and recovery.

Layer 6 – Presentation layer

The penultimate layer is the presentation layer, which takes care of syntax and semantics between applications, through dealing with complex data representation. It allows the last layer to be independent from the encoding used to represent the data. The presentation of the OSI model used ASN.1 encoding, but there is a wide range of different presentation protocols that are widely used, such as XML and JSON.

Layer 7 – Application layer

The last layer, the application layer, is the one that communicates with the application directly. The application is not considered as a part of the OSI model and the layer is responsible for defining the interfaces used by the applications. It includes protocols such as FTP, DNS, and SMTP.

TCP/IP – Internet protocol suite

The **Transmission Control Protocol/Internet Protocol** (**TCP/IP**), or internet protocol suite, is another model composed by fewer layers than the OSI, and it is a model that has been widely adopted.

Layer 1 – Link layer

The first layer is the link layer, a combination of OSI's physical and data links, and it defines how the local networking communications will happen by specifying a protocol, such as MAC (which includes Ethernet and Wi-Fi).

Layer 2 – Internet layer

The internet layer is the second layer, and it can be compared to the OSI's network. It defines a common interface that allows different networks to communicate effectively without any awareness of the underlying topology of each one of them. This layer is responsible for communications between the nodes in a **Local Area Network** (**LAN**) and the global interconnected networks that compose the internet.

Layer 3 – Transport layer

The third layer is similar to the fourth OSI layer. It handles the end-to-end communication of two devices and it also takes care of the error check and recovery, leaving the upper layer unaware of the complexity of the data. It defines two main protocols—TCP, which allows the receiver to get the data in the correct sequence by using an acknowledgement system, and a **User Data Protocol** (**UDP**), which does not apply error control or acknowledgement from the receiver.

Layer 4 – Application layer

The last layer, the application layer, sums up the last three levels of OSI—session, presentation, and application. This layer defines the architecture used by the applications, such as peer-to-peer or client and server, and the protocols used by the applications, such as SSH, HTTP, or SMTP. Each process is an address with a virtual endpoint of communication called *port*.

Understanding socket programming

The Go standard library allows us to easily interact with the transport layer, using both TCP and UDP connections. In this section, we will look at how to expose a service using a socket and how to look it up and use it from another application.

Network package

The tools needed to create and handle a TCP connection are located inside the net package. The main interface of the package is Conn, which represents a connection.

It has four implementations:

- `IPConn`: Raw connection that uses the IP protocol, the one that TCP and UDP connection are built on
- `TCPConn`: An IP connection that uses the TCP protocol
- `UDPConn`: An IP connection that uses the UDP protocol
- `UnixConn`: A Unix domain socket, where the connection is meant for processes on the same machine

In the following sections, we are going to look at how to use both TCP and UDP differently, and also how to use an IPConn to implement a custom implementation of a communication protocol.

TCP connections

The TCP is the protocol that is most used on the internet, and it enables delivery of data (bytes) that are ordered. The main focus of this protocol is reliability, which is obtained by establishing a two-way communication, where the receiver sends an acknowledgement signal each time it receives a datagram successfully.

A new connection can be created using the `net.Dial` function. It is a generic function that accepts different networks, such as the following:

- `tcp`, `tcp4` (IPv4-only), `tcp6` (IPv6-only)
- `udp`, `udp4` (IPv4-only), `udp6` (IPv6-only)
- `ip`, `ip4` (IPv4-only), `ip6` (IPv6-only)
- `unix` (socket stream), `unixgram` (socket datagram), and `unixpacket` (socket packet)

A TCP connection can be created, specifying the `tcp` protocol, together with the host and port:

```
conn, err := net.Dial("tcp", "localhost:8080")
```

A more direct way of creating a connection is `net.DialTCP`, which allows you to specify both a local and remote address. It requires you to create a `net.TCPAddr` in order to use it:

```
addr, err := net.ResolveTCPAddr("tcp", "localhost:8080")
if err != nil {
    // handle error
}
conn, err := net.DialTCP("tcp", nil, addr)
```

```
if err != nil {
    // handle error
}
```

In order to receive and handle connections, there is another interface, net.Listener, which has four different implementations—one per connection type. For connections, there is a generic net.Listen function and a specific net.ListenTCP function.

We can try to build a simple application that creates a TCP listener and connects to it, sending whatever comes from the standard input. The app should create a listener to start a connection in the background, which will send standard input to the connection, then accept it, and handle it. We will use the newline character as the delimiter for messages. This is shown in the following code:

```
func main() {
    if len(os.Args) != 2 {
        log.Fatalln("Please specify an address.")
    }
    addr, err := net.ResolveTCPAddr("tcp", os.Args[1])
    if err != nil {
        log.Fatalln("Invalid address:", os.Args[1], err)
    }
    listener, err := net.ListenTCP("tcp", addr)
    if err != nil {
        log.Fatalln("Listener:", os.Args[1], err)
    }
    log.Println("<- Listening on", addr)

    go createConn(addr)

    conn, err := listener.AcceptTCP()
    if err != nil {
        log.Fatalln("<- Accept:", os.Args[1], err)
    }
    handleConn(conn)
}
```

The connection creation is pretty simple. It creates the connection and reads messages from the standard input and forwards them to the connection by writing them:

```
func createConn(addr *net.TCPAddr) {
    defer log.Println("-> Closing")
    conn, err := net.DialTCP("tcp", nil, addr)
    if err != nil {
        log.Fatalln("-> Connection:", err)
    }
    log.Println("-> Connection to", addr)
```

```
    r := bufio.NewReader(os.Stdin)
    for {
        fmt.Print("# ")
        msg, err := r.ReadBytes('\n')
        if err != nil {
            log.Println("-> Message error:", err)
        }
        if _, err := conn.Write(msg); err != nil {
            log.Println("-> Connection:", err)
            return
        }
    }
}
```

In our use case, the connection that sends data will be closed with a special message, \q, which will be interpreted as a command. Accepting a connection in the listener creates another connection that represents the ones obtained by the dialing operation. The connection created by the listener will be receiving the messages from the dialing one and acting accordingly. It will interpret a special message, such as \q, with a specific action; otherwise, it will just print the message on screen, as shown in the following code:

```
func handleConn(conn net.Conn) {
    r := bufio.NewReader(conn)
    time.Sleep(time.Second / 2)
    for {
        msg, err := r.ReadString('\n')
        if err != nil {
            log.Println("<- Message error:", err)
            continue
        }
        switch msg = strings.TrimSpace(msg); msg {
        case `\q`:
            log.Println("Exiting...")
            if err := conn.Close(); err != nil {
                log.Println("<- Close:", err)
            }
            time.Sleep(time.Second / 2)
            return
        case `\x`:
            log.Println("<- Special message `\\x` received!")
        default:
            log.Println("<- Message Received:", msg)
        }
    }
}
```

The following code example creates both a client and server in one application, but it could be easily split into two apps—a server (capable of handling more connections at once) and a client, creating a single connection to the server. The server will have an `Accept` loop that handles the received connection on a separate goroutine. The `handleConn` function is the same as we defined earlier:

```go
func main() {
    if len(os.Args) != 2 {
        log.Fatalln("Please specify an address.")
    }
    addr, err := net.ResolveTCPAddr("tcp", os.Args[1])
    if err != nil {
        log.Fatalln("Invalid address:", os.Args[1], err)
    }
    listener, err := net.ListenTCP("tcp", addr)
    if err != nil {
        log.Fatalln("Listener:", os.Args[1], err)
    }
    for {
        time.Sleep(time.Millisecond * 100)
        conn, err := listener.AcceptTCP()
        if err != nil {
            log.Fatalln("<- Accept:", os.Args[1], err)
        }
        go handleConn(conn)
    }
}
```

The client will create the connection and send messages. `createConn` will be the same as we defined earlier:

```go
func main() {
    if len(os.Args) != 2 {
        log.Fatalln("Please specify an address.")
    }
    addr, err := net.ResolveTCPAddr("tcp", os.Args[1])
    if err != nil {
        log.Fatalln("Invalid address:", os.Args[1], err)
    }
    createConn(addr)
}
```

In the separated client and server, it's possible to test what happens when the client or server closes the connection.

UDP connections

UDP is another protocol that is widely used on the internet. It focuses on low latency, which is why it is not as reliable as TCP. It has many applications, from online gaming, to media streaming, and **Voice over Internet Protocol (VoIP)**. In UDP, if a packet is not received, it is lost and will not be sent back again as it would be in TCP. Imagine a VoIP call, where if there's a connection problem you will lose part of the conversation, but when you resume, you keep communicating almost in real time. Using TCP for this kind of application could result in latency accumulating for every packet loss—making the conversation impossible.

In the following example, we'll create a client and a server application. The server will be some sort of echo, sending back the message received from the client, but it will also reverse the message content.

The client will be pretty similar to the TCP one, with some exceptions—it will use the net.ResolveUDPAddr function to get the address and it will use net.DialUDP to get the connection:

```go
func main() {
    if len(os.Args) != 2 {
        log.Fatalln("Please specify an address.")
    }
    addr, err := net.ResolveUDPAddr("udp", os.Args[1])
    if err != nil {
        log.Fatalln("Invalid address:", os.Args[1], err)
    }
    conn, err := net.DialUDP("udp", nil, addr)
    if err != nil {
        log.Fatalln("-> Connection:", err)
    }
    log.Println("-> Connection to", addr)
    r := bufio.NewReader(os.Stdin)
    b := make([]byte, 1024)
    for {
        fmt.Print("# ")
        msg, err := r.ReadBytes('\n')
        if err != nil {
            log.Println("-> Message error:", err)
        }
        if _, err := conn.Write(msg); err != nil {
            log.Println("-> Connection:", err)
            return
        }
        n, err := conn.Read(b)
        if err != nil {
```

```
                log.Println("<- Receive error:", err)
            }
        msg = bytes.TrimSpace(b[:n])
        log.Printf("<- %q", msg)
        }
    }
```

The server will be pretty different from the TCP one. The main difference is that with TCP, we have a listener that accepts different connections, which are handled separately; meanwhile, the UDP listener is a connection. It can receive data blindly or use the `ReceiveFrom` method that will also return the recipient's address. This can be used in the `WriteTo` method to answer, as shown in the following code:

```
func main() {
    if len(os.Args) != 2 {
        log.Fatalln("Please specify an address.")
    }
    addr, err := net.ResolveUDPAddr("udp", os.Args[1])
    if err != nil {
        log.Fatalln("Invalid address:", os.Args[1], err)
    }
    conn, err := net.ListenUDP("udp", addr)
    if err != nil {
        log.Fatalln("Listener:", os.Args[1], err)
    }

    b := make([]byte, 1024)
    for {
        n, addr, err := conn.ReadFromUDP(b)
        if err != nil {
            log.Println("<-", addr, "Message error:", err)
            continue
        }
        msg := bytes.TrimSpace(b[:n])
        log.Printf("<- %q from %s", msg, addr)
        for i, l := 0, len(msg); i < l/2; i++ {
            msg[i], msg[l-1-i] = msg[l-1-i], msg[i]
        }
        msg = append(msg, '\n')
        if _, err := conn.WriteTo(b[:n], addr); err != nil {
            log.Println("->", addr, "Send error:", err)
        }
    }
}
```

Encoding and checksum

It's a good practice to set some form of encoding between the clients and the server, and even better practice if the encoding includes a checksum to verify data integrity. We could improve the example from the last section with a custom protocol that does both encoding and checksum. Let's start by defining the encoding function, where a given message will return the following byte sequence:

Function	Byte sequence
The first four bytes will follow a sequence	2A 00 2A 00
Two bytes will be the message length stored using Little Endian order (least significant byte first)	08 00
Four bytes for data checksum	00 00 00 00
Followed by the raw message	0F 1D 3A FF ...
Closing with the same starting sequence	2A 00 2A 00

The Checksum function will be calculated by summing the message content, using a group of five bytes in Little Endian (least significant byte first), adding any spare byte left one by one, and then taking the first four bytes of the sum as a Little Endian:

```
func Checksum(b []byte) []byte {
    var sum uint64
    for len(b) >= 5 {
        for i := range b[:5] {
            v := uint64(b[i])
            for j := 0; j < i; j++ {
                v = v * 256
            }
            sum += v
        }
        b = b[5:]
    }
    for _, v := range b {
        sum += uint64(v)
    }
    s := make([]byte, 8)
    binary.LittleEndian.PutUint64(s, sum)
    return s[:4]
}
```

Now, let's create a function that will encapsulate the message using the protocol we defined:

```
var ErrLength = errors.New("message too long")

func CreateMessage(content []byte) ([]byte, error) {
    if len(content) > 65535 {
        return nil, ErrLength
    }
    data := make([]byte, 0, len(content)+14)
    data = append(data, Sequence...)
    data = append(data, byte(len(content)/256), byte(len(content)%256))
    data = append(data, Checksum(content)...)
    data = append(data, content...)
    data = append(data, Sequence...)
    return data, nil
}
```

We also need another function that does the opposite, checking whether a message is valid and extracting its content:

```
func MessageContent(b []byte) ([]byte, error) {
    n := len(b)
    if n < 14 {
        return nil, fmt.Errorf("Too short")
    }
    if open := b[:4]; !bytes.Equal(open, Sequence) {
        return nil, fmt.Errorf("Wrong opening sequence %x", open)
    }
    if length := int(b[4])*256 + int(b[5]); n-14 != length {
        return nil, fmt.Errorf("Wrong length: %d (expected %d)", length,
n-14)
    }
    if close := b[n-4 : n]; !bytes.Equal(close, Sequence) {
        return nil, fmt.Errorf("Wrong closing sequence %x", close)
    }
    content := b[10 : n-4]
    if !bytes.Equal(Checksum(content), b[6:10]) {
        return nil, fmt.Errorf("Wrong checksum")
    }
    return content, nil
}
```

We can now use them for encoding and decoding messages. For instance, we could improve the UDP client and server from the previous section and we could encode when sending:

```
// Send
data, err := common.CreateMessage(msg)
if err != nil {
    log.Println("->", addr, "Encode error:", err)
    continue
}
if _, err := conn.WriteTo(data, addr); err != nil {
    log.Println("->", addr, "Send error:", err)
}
```

And we can decode the bytes received for incoming messages in order to extract the content:

```
//Receive
n, addr, err := conn.ReadFromUDP(b)
if err != nil {
    log.Println("<-", addr, "Message error:", err)
    continue
}
msg, err := common.MessageContent(b[:n])
if err != nil {
    log.Println("<-", addr, "Decode error:", err)
    continue
}
log.Printf("<- %q from %s", msg, addr)
```

In order to verify that the content we received is valid, we are using the `MessageContent` utility function defined previously. This will check for headers, length, and checksum. It will only extract the bytes that compose the message.

Web servers in Go

One of Go's biggest and most successful applications is the creation of web servers. In this section, we will see what a web server actually is, how the HTTP protocol works, and how to implement a web server application using both the standard library and third-party packages.

Web server

A web server application is software that can serve content using the HTTP protocol (and some other related ones) over a TCP/IP network. There are many well-known web server applications, such as Apache, NGINX, and Microsoft IIS. Common server use case scenarios include the following:

- **Serving static files, such websites and related resources**: HTML pages, images, style sheets, and scripts.
- **Exposing a web application**: An application that runs in the server with an HTML-based interface, which requires a browser to access it.
- **Exposing a web API**: Remote interfaces that are not used by the user but from other applications. Refer to `Chapter 1`, *An Introduction to System Programming*, for more details.

HTTP protocol

The HTTP protocol is the cornerstone of a web server. Its design started in 1989. The main usage of HTTP is the request and response paradigm, where the client sends a request and the server returns back a response to the client.

Uniform Resource Locators (**URLs**) are unique identifiers for an HTTP request, and they have the following structure:

Part	Example
Protocol	`http`
`://`	`://`
Host	`www.website.com`
Path	`/path/to/some-resource`
?	?
Query (optional)	`query=string&with=values`

From the preceding table, we can derive the following:

- There are several different protocols beside HTTP and its encrypted version (HTTPS), such as the **File Transfer Protocol** (**FTP**) and its secure counterpart, the **SSH File Transfer Protocol** (**SFTP**).
- The host could either be an actual IP or a hostname. When a hostname is selected, there is another player, a **Domain Name Server** (**DNS**), that acts as a phonebook between hostname and physical addresses. The DNS translates hostnames to IPs.

- The path is the resource desired in the server and it is always absolute.
- The query string is something added to a path after a question mark. It is a series of key value pairs in the form `key=value` and they are separated by an `&` sign.

HTTP is a textual protocol and it contains some of the elements of the URL and some other information—method, headers, and body.

The request body is the information sent to server, such as form values, or uploaded files.

Headers are metadata relative to the request, one per line, in a `Key: Value; extra data` form. There is a list of defined headers with specific functions, such as `Authorization`, `User-Agent`, and `Content-Type`.

Some methods express the action to execute on a resource. These are the most commonly used methods:

- `GET`: A representation of the selected resource
- `HEAD`: Similar to `GET`, but without any response body
- `POST`: Submits a resource to the server, usually a new one
- `PUT`: Submits a new version of a resource
- `DELETE`: Removes a resource
- `PATCH`: Requests specific change to a resource

This is what an HTTP request would look like:

```
POST /resource/ HTTP/1.1
User-Agent: Mozilla/4.0 (compatible; MSIE5.01; Windows NT)
Host: www.website.com
Content-Length: 1024
Accept-Language: en-us

the actual request content
that is optional
```

From the preceding code, we can see the following:

- The first line is a space separated triplet: method—path—protocol.
- It's followed by one line per header.
- One empty line as separator.
- The optional request body.

For each request, there is a response that has a structure that is pretty similar to an HTTP request. The only part that differs is the first line that contains a different space-separated triplet: HTTP version—status code—reason.

The status code is an integer that represents the outcome of the request. There are four main status categories:

- 100: Information/request was received and will have further processing
- 200: A successful request; for instance, OK 200 or Created 201
- 300: A redirection to another URL, temporary or permanent
- 400: A client-side error, such as Not Found 404 or Conflict 409
- 500: A server-side error, such as Internal Server Error 503

This is a what an HTTP response would look like:

```
HTTP/1.1 200 OK
Content-Length: 88
Content-Type: text/html

<html>
  <body>
    <h1>Sample Page</h1>
  </body>
</html>
```

HTTP/2 and Go

The most commonly used version of HTTP is HTTP/1.1 , dated 1997. In 2009, Google started a new project to create a faster successor to HTTP/1.1, named SPDY. This protocol eventually became what is now a version 2.0 of the **Hypertext Transfer Protocol, HTTP/2**.

It is built in a way that existing web applications work, but there are new features for applications that are using the new protocol, including a faster communication speed. Some of the differences include the following:

- It is binary (HTTP/1.1 is textual).
- It is fully multiplexed, and can request data in parallel, using one TCP connection.
- It uses header compression to reduce overhead.

- Servers can push responses to the client, instead of being asked by clients periodically.
- It has a faster protocol negotiation—thanks to the **Application Layer Protocol Negotiation (ALPN)** extension.

HTTP/2 is supported by all major modern browsers. Go version 1.6 included transparent support for HTTP/2 and version 1.8 introduced the ability to push responses from the server to the clients.

Using the standard package

We will now see how to create a web server in Go, using the standard package. Everything is contained in the net/http package, which exposes a series of functions for making HTTP requests and creating HTTP servers.

Making a HTTP request

The package exposes a http.Client type that can be used to make requests. If the requests are simple GET or POST, then there are dedicated methods. The package also offers a function with the same name, but it's just a shorthand to the respective methods for the DefaultClient instance. Check the following code:

```
resp, err := http.Get("http://example.com/")
resp, err := client.Get("http://example.com/")
...
resp, err := http.Post("http://example.com/upload", "image/jpeg", &buf)
resp, err := client.Post("http://example.com/upload", "image/jpeg", &buf)
...
values := url.Values{"key": {"Value"}, "id": {"123"}}
resp, err := http.PostForm("http://example.com/form", values)
resp, err := client.PostForm("http://example.com/form", values)
```

For any other kind of requirement, the Do method allows us to execute a specific http.Request. The NewRequest function allows us to specify any io.Reader:

```
req, err := http.NewRequest("GET", "http://example.com", nil)
// ...
req.Header.Add("Content-Type", "text/html")
resp, err := client.Do(req)
// ...
```

`http.Client` has several fields and many of these are interfaces that allows us to use the default implementation or a custom one. The first is `CookieJar` that allows the client to store and reuse web cookies. A cookie is a piece of data that the browser sends to the client and the client can send back to the server to replace headers, such as authentication. The default client does not use a cookie jar. The other interface is `RoundTripper`, which has only one method, `RoundTrip`, which gets a request and returns a response. There is a `DeafultTransport` value used if no value is specified that can also be used to compose a custom implementation of `RoundTripper`. `http.Response` returned by the client has also a body, which is `io.ReadCloser`, and its closure is the responsibility of the application. That's why it's recommended to use a deferred `Close` statement as soon as the response is obtained. In the following example, we will implement a custom transport that logs the requested URL and modifies one header before executing the standard round tripper:

```
type logTripper struct {
    http.RoundTripper
}

func (1 logTripper) RoundTrip(r *http.Request) (*http.Response,
    error) {
        log.Println(r.URL)
        r.Header.Set("X-Log-Time", time.Now().String())
        return l.RoundTripper.RoundTrip(r)
}
```

We will use this transport in a client that we will use to make a simple request:

```
func main() {
    client := http.Client{Transport: logTripper{http.DefaultTransport}}
    req, err := http.NewRequest("GET",
"https://www.google.com/search?q=golang+net+http", nil)
    if err != nil {
        log.Fatal(err)
    }
    resp, err := client.Do(req)
    if err != nil {
        log.Fatal(err)
    }
    defer resp.Body.Close()
    log.Println("Status code:", resp.StatusCode)
}
```

Creating a simple server

The other functionality offered by the package is the server creation. The main interface of the package is Handle, which has one method, ServeHTTP ,that uses the request to write a response. Its simplest implementation is HandlerFunc, which is a function with the same signature of ServeHTTP and implements Handler by executing itself.

The ListenAndServe function starts a HTTP server using the given address and handler. If no handler is specified, it uses the DefaultServeMux variable. ServeMux is special type of Handler that manages the execution of different handlers, depending on the URL path requested. It has two methods, Handle and HandleFunc, that allow the user to specify the path and the respective handler. The package also offers generic handler functions that are similar to what we have seen for the Client, they will call the methods with same name from the default ServerMux.

In the following example, we will create a customHandler and create a simple server with some endpoints, including the custom one:

```go
type customHandler int

func (c *customHandler) ServeHTTP(w http.ResponseWriter, r
    *http.Request) {
        fmt.Fprintf(w, "%d", *c)
        *c++
}

func main() {
    mux := http.NewServeMux()
    mux.HandleFunc("/hello", func(w http.ResponseWriter, r
        *http.Request) {
            fmt.Fprintf(w, "Hello!")
    })
    mux.HandleFunc("/bye", func(w http.ResponseWriter, r
        *http.Request) {
            fmt.Fprintf(w, "Goodbye!")
    })
    mux.HandleFunc("/error", func(w http.ResponseWriter, r
        *http.Request) {
            w.WriteHeader(http.StatusInternalServerError)
            fmt.Fprintf(w, "An error occurred!")
    })
    mux.Handle("/custom", new(customHandler))
    if err := http.ListenAndServe(":3000", mux); err != nil {
        log.Fatal(err)
    }
}
```

Serving filesystem

The Go standard package allows us to easily serve a certain directory in the filesystem, using the net.FileServer function, which, when given the net.FileSystem interface, returns a Handler that serves that directory. The default implementation is net.Dir, which is a custom string that represents a directory in the system. The FileServer function already has a protection mechanism that prevents us to use a relative path, such as../../../dir, to access directories outside the served one.

The following is a sample file server that uses the directory provided as an argument, as the root of the file served:

```
func main() {
    if len(os.Args) != 2 {
        log.Fatalln("Please specify a directory")
    }
    s, err := os.Stat(os.Args[1])
    if err == nil && !s.IsDir() {
        err = errors.New("not a directory")
    }
    if err != nil {
        log.Fatalln("Invalid path:", err)
    }
    http.Handle("/", http.FileServer(http.Dir(os.Args[1])))
    if err := http.ListenAndServe(":3000", nil); err != nil {
        log.Fatal(err)
    }
}
```

Navigating through routes and methods

The HTTP method used is stored in the Request.Method field. This field can be used inside a handler to have different behaviors for each supported method:

```
switch r.Method {
case http.MethodGet:
    // GET implementation
case http.MethodPost:
    // POST implementation
default:
    http.NotFound(w, r)
}
```

The advantage of the `http.Handler` interface is that we can define our custom types. This can make the code more readable and can generalize this method-specific behavior:

```
type methodHandler map[string]http.Handler

func (m methodHandler) ServeHTTP(w http.ResponseWriter, r
        *http.Request) {
            h, ok := m[strings.ToUpper(r.Method)]
            if !ok {
                http.NotFound(w, r)
                return
            }
    h.ServeHTTP(w, r)
}
```

This will make the code much more readable, and it will be reusable for different paths:

```
func main() {
    http.HandleFunc("/path1", methodHandler{
        http.MethodGet: http.HandlerFunc(func(w http.ResponseWriter, r
*http.Request) {
            fmt.Fprint(w, "Showing record")
        }),
        http.MethodPost: http.HandlerFunc(func(w http.ResponseWriter, r
*http.Request) {
            fmt.Fprint(w, "Updated record")
        }),
    })
    if err := http.ListenAndServe(":3000", nil); err != nil {
        log.Fatal(err)
    }
}
```

Multipart request and files

The request body is an `io.ReadCloser`. This means that it is the server's responsibility to close it. For a file upload, the body is not content of the file directly, but it is usually a multipart request, which is a request that specifies a boundary in the header and uses it inside the body to separate the message in parts.

This is a sample multipart message:

```
MIME-Version: 1.0
Content-Type: multipart/mixed; boundary=xxxx

This part before boundary is ignored
```

```
--xxxx
Content-Type: text/plain

First part of the message. The next part is binary data encoded in base64
--xxxx
Content-Type: application/octet-stream
Content-Transfer-Encoding: base64

PGh0bWw+CiAgPGh1YWQ+CiAgPC9oZWFkPgogIDxib2R5PgogICAgPHA+VGhpcyBpcyB0aGUg
Ym9keSBvZiB0aGUgbWVzc2FnZS48L3A+CiAgPC9ib2R5Pgo8L2h0bWw+Cg==
--xxxx--
```

We can see that the boundary has two dashes as a prefix and it's followed by a newline, and the final boundary also has two dashes as a suffix. In the following example, the server will handle a file upload, with a small form to send the request from a browser.

Let's define some constants that we will use in the handler:

```
const (
    param = "file"
    endpoint = "/upload"
    content = `<html><body>` +
        `<form enctype="multipart/form-data" action="%s" method="POST">` +
        `<input type="file" name="%s"/><input type="submit"
value="Upload"/>` +
        `</form></html></body>`
)
```

Now, we can define the handler function. The first part should show the template if the method is GET, as it executes the upload on POST and returns a not found status otherwise:

```
mux.HandleFunc(endpoint, func(w http.ResponseWriter, r
    *http.Request) {
        if r.Method == "GET" {
            fmt.Fprintf(w, content, endpoint, param)
            return
        } else if r.Method != "POST" {
            http.NotFound(w, r)
            return
        }

    path, err := upload(r)
    if err != nil {
        http.Error(w, err.Error(), http.StatusInternalServerError)
        return
    }
    fmt.Fprintf(w, "Uploaded to %s", path)
})
```

The `upload` function will use the `Request.FormFile` method that returns the file and its metadata:

```
func upload(r *http.Request) (string, error) {
    f, h, err := r.FormFile(param)
    if err != nil {
        return "", err
    }
    defer f.Close()

    p := filepath.Join(os.TempDir(), h.Filename)
    fw, err := os.OpenFile(p, os.O_WRONLY|os.O_CREATE, 0666)
    if err != nil {
        return "", err
    }
    defer fw.Close()

    if _, err = io.Copy(fw, f); err != nil {
        return "", err
    }
    return p, nil
}
```

HTTPS

If you want your web server to use HTTPS instead of relying on external applications, such as NGINX, you can easily do so if you already have a valid certificate. If you don't, you can use OpenSSL to create one:

```
> openssl genrsa -out server.key 2048

> openssl req -new -x509 -sha256 -key server.key -out server.crt -days 3650
```

The first command generates a private key, while the second one creates a public certificate that you need for the server. The second command will also require a lot of additional information in order to create the certificate, from country name to email address.

Once everything is ready, in order to create an HTTPS server, the `http.ListenAndServe` function needs to be replaced with its secure counterpart: `http.ListenAndServeTLS`:

```
func main() {
    http.HandleFunc("/hello", func(w http.ResponseWriter, r
        *http.Request) {
            fmt.Fprint(w, "Hello!")
    })
```

```
    err := http.ListenAndServeTLS(":443", "server.crt", "server.key", nil)
    if err != nil {
        log.Fatal("ListenAndServe: ", err)
    }
}
```

Third-party packages

The Go open source community develops a lot of packages that integrate with net/http, implementing the Handler interface, but offer a set of unique capabilities that allow for an easier development of web servers.

gorilla/mux

The github.com/gorilla/mux package contains another implementation of Handler that enhances the capabilities of the standard ServeMux:

- Better URL matching to handlers using any element from the URL, including schema, method, host, or query values.
- URL elements, such as host, paths, and query keys can have placeholders (which can also use regular expressions).
- Routes can be defined hierarchically with the help of subroutes, nested routes which test a part of the path.
- Handlers can be also used as middleware, before the main handler, for all paths or for a subset of paths.

Let's start with an example of matching using other path elements:

```
r := mux.NewRouter()
// only local requests
r.Host("localhost:3000")
// only when some header is present
r.Headers("X-Requested-With", "XMLHttpRequest")
only when a query parameter is specified
r.Queries("access_key", "0x20")
```

Variables are another really useful feature, that allow us to specify placeholders, and get their values with the auxiliary function, `mux.Vars`, as shown in the following example:

```
r := mux.NewRouter()
r.HandleFunc("/products/", ProductsHandler)
r.HandleFunc("/products/{key}/", ProductHandler)
r.HandleFunc("/products/{key}/details", ProductDetailsHandler)
...
// inside an handler
vars := mux.Vars(request)
key:= vars["key"]
```

`Subrouter` is another helpful function to group same prefixes routes. This allows us to simplify the previous code to the following code snippet:

```
r := mux.NewRouter()
s := r.PathPrefix("/products").Subrouter()
s.HandleFunc("/", ProductsHandler)
s.HandleFunc("/{key}/", ProductHandler)
s.HandleFunc("/{key}/details", ProductDetailsHandler)
```

Middleware is also very useful when combined with subroutes, to execute some common tasks as authentication and verification:

```
r := mux.NewRouter()
pub := r.PathPrefix("/public").Subrouter()
pub.HandleFunc("/login", LoginHandler)
priv := r.PathPrefix("/private").Subrouter()
priv.Use(AuthHandler)
priv.HandleFunc("/profile", ProfileHandler)
priv.HandleFunc("/logout", LogoutHandler)
```

gin-gonic/gin

The `github.com/gin-gonic/gin` package is another Go web framework that extends the capabilities of a Go HTTP server with many shorthand and auxiliary functions. Its features include the following:

- **Speed**: Its routing is fast and has very little memory footprint.
- **Middleware**: It allows to define and use intermediate handlers with a full control of their flow.
- **Panic-free**: It comes with middleware that recovers from panics.
- **Grouping**: It can group routes with the same prefix together.

- **Errors**: It manages and collects error that happen during the request.
- **Rendering**: It comes out of the box with renderers for most web formats (JSON, XML, HTML).

The core of the package is `gin.Engine`, which is also a `http.Handler`. The `gin.Default` function returns an engine that uses two middlewares—`Logger`, which prints the result of each HTTP request received, and `Recovery`, which recovers from panics. The other option is to use the `gin.New` function, which returns an engine with no middleware.

It allows us to bind a handler to a single HTTP method with a series of the engine's methods named after their HTTP counterpart:

- DELETE
- GET
- HEAD
- OPTIONS
- PATCH
- POST
- PUT
- Any (catch all for any HTTP method)

There's also a `group` method that returns a route grouping for the selected path, which exposes all of the preceding methods:

```
router := gin.Default()

router.GET("/resource", getResource)
router.POST("/resource", createResource)
router.PUT("/resource", updateResoure)
router.DELETE("/resource", deleteResource)
// with use grouping
g := router.Group("/resource")
g.GET("", getResource)
g.POST("", createResource)
g.PUT("", updateResoure)
g.DELETE("", deleteResource)
```

The handlers in this framework have a different signature. Instead of having a response writer and a request as arguments, it uses `gin.Context`, a structure that wraps both, and offers many shorthands and utilities. For instance, the package offers the possibility of using placeholders in the URL and the context enables these parameters to be read:

```
router := gin.Default()
router.GET("/hello/:name", func(c *gin.Context) {
    c.String(http.StatusOK, "Hello %s!", c.Param("name"))
})
```

We can also see in the example that the context offers a `String` method that enables us to write an HTTP status and the content of the response with a one-liner.

Other functionalities

There are additional features for web servers. Some of them are already supported by the standard library (such as HTTP/2 Pusher) and others are available with experimental packages or third-party libraries (such as WebSockets).

HTTP/2 Pusher

We have already discussed that Golang supports the HTTP/2 server side push functionality since version 1.8. Let's see how to use it in an application. Its usage is pretty simple; if the request can be casted to the `http.Pusher` interface, it can be used to push additional requests in the main interface. In this example, we use it to sideload a SVG image, together with the page:

```
func main() {
    const imgPath = "/image.svg"
    http.HandleFunc("/", func(w http.ResponseWriter, r
        *http.Request) {
            pusher, ok := w.(http.Pusher)
            if ok {
                fmt.Println("Push /image")
                pusher.Push(imgPath, nil)
            }
        w.Header().Add("Content-Type", "text/html")
        fmt.Fprintf(w, `<html><body><img src="%s"/>`+
            `</body></html>`, imgPath)
    })
    http.HandleFunc(imgPath, func(w http.ResponseWriter, r
        *http.Request) {
            w.Header().Add("Content-Type", "image/svg+xml")
```

```
                fmt.Fprint(w, `<?xml version="1.0" standalone="no"?>
<svg xmlns="http://www.w3.org/2000/svg">
  <rect width="150" height="150" style="fill:blue"/>
</svg>`)
        })
        if err := http.ListenAndServe(":3000", nil); err != nil {
            fmt.Println(err)
        }
}
```

This will result in two separate requests for HTTP/1, and one single request for HTTP/2, where the second request is obtained using the push capabilities of the browser.

WebSockets protocol

The HTTP protocol achieves only one-way communication, while the WebSocket protocol is a full duplex communication between client and server. The Go experimental library offers support for WebSocket with the `golang.org/x/net/websocket` package, and Gorilla has another implementation with its own `github.com/gorilla/websocket`.

The second one is far more complete, and it's used in the `github.com/olahol/melody` package, which implements a framework for an easy WebSocket communication. Each package offers different working examples on the WebSocket server and client pair.

Beginning with the template engine

Another very powerful tool is the Go templating engine, available in `text/template`. Its functionalities are replicated and extended in the `html/template` package, which constitutes another powerful tool for web development with Go.

Syntax and basic usage

The template package enables us to separate presentation from data, using text files and data structures. The template engine defines two delimiters—left and right—for opening and closing actions that represent data evaluation. The default delimiters are `{{` and `}}`, and the template evaluates only what's included within these delimiters, leaving the rest untouched.

Usually the data bound to the template is a structure or a map and it is accessible anywhere in the template with the $ variable. Whether it's a map or a struct, the fields are always accessed in the same way, using the .Field syntax. If the dollar is omitted, the value is referred to the current context, which is $ if it's not in special statements, such as loops. Outside of these exceptions, the {{$.Field}} and {{.Field}} statements are equivalent.

The flow in a template is controlled with a condition statement, {{if}}, and a loop statement, {{range}}, and both terminate with the {{end}} statement. The condition statement also offers the possibility of a chain {{else if}} statement to specify another condition, acting like a switch, and an {{else}} statement, which can be considered the default of a switch. {{else}} can be used with the range statement and it's executed when the argument of the range is nil or has zero length.

Creating, parsing, and executing templates

The template.Template type is a collector of one or more templates and can be initialized in several ways. The template.New function creates a new empty template with the given name, which can be used to call the Parse method that uses a string to create a template. Consider the following code:

```
var data = struct {
    Question string
    Answer int
}{
    Question: "Answer to the Ultimate Question of Life, " +
        "the Universe, and Everything",
    Answer: 42,
}
tpl, err := template.New("question-answer").Parse(`
    <p>Question: {{.Question}}</p>
    <p>Answer: {{.Answer}}</p>
`)
if err != nil {
    log.Fatalln("Error:", err)
}
if err = tpl.Execute(os.Stdout, data); err != nil {
    log.Fatalln("Error:", err)
}
```

The full example is available here: `https://play.golang.org/p/k-t0Ns1b2Mv`.

Templates can also be loaded and parsed from a filesystem, using `template.ParseFiles`, which takes a list of files, and `template.ParseGlob`, which uses the `glob` Unix command syntax to select a list of files. Let's create a template file with the following content:

```
<html>
    <body>
        <h1>{{.name}}</h1>
        <ul>
            <li>First appearance: {{.appearance}}</li>
            <li>Style: {{.style}}</li>
        </ul>
    </body>
</html>
```

We can use one of these two functions to load it and execute it with some sample data:

```
func main() {
    tpl, err := template.ParseGlob("ch9/template/parse/*.html")
    if err != nil {
        log.Fatal("Error:", err)
    }
    data := map[string]string{
        "name": "Jin Kazama",
        "style": "Karate",
        "appearance": "Tekken 3",
    }
    if err := tpl.Execute(os.Stdout, data); err != nil {
        log.Fatal("Error:", err)
    }
}
```

When multiple templates are loaded, the `Execute` method will use the last one. If a specific template needs to be selected, there is another method, `ExecuteTemplate`, which also receives the template name as an argument, to specify which template to use.

Conditions and loops

The `range` statement can be used in different ways—the simplest way is just calling `range` followed by the slice or map that you want to iterate.

Alternatively, you can specify the values, or the index and the value:

```
var a = []int{1, 2, 3, 4}
`{{ range . }} {{.}} {{ end }}` // simple
`{{ range $v := . }} {{$v}} {{ end }}` // value
`{{ range $i, $v := . }} {{$v}} {{ end }}` // index and value
```

When in a loop, the `{{.}}` variable assumes the value of the current element in the iteration. The following example loops a slice of items:

```
var data = []struct {
    Question, Answer string
}{{
    Question: "Answer to the Ultimate Question of Life, " +
        "the Universe, and Everything",
    Answer: "42",
}, {
    Question: "Who you gonna call?",
    Answer: "Ghostbusters",
}}
tpl, err := template.New("question-answer").Parse(`{{range .}}
Question: {{.Question}}
Answer: {{.Answer}}
{{end}}`)
if err != nil {
    log.Fatalln("Error:", err)
}
if err = tpl.Execute(os.Stdout, data); err != nil {
    log.Fatalln("Error:", err)
}
```

The full example is available here: https://play.golang.org/p/MtU_d9CsFb-.

The next example is a use case of conditional statements that have also made use of the `lt` function:

```
var data = []struct {
    Name string
    Score int
}{
    {"Michelangelo", 30},
    {"Donatello", 50},
    {"Leonardo", 80},
    {"Raffaello", 100},
}
tpl, err := template.New("question-answer").Parse(`{{range .}}
{{.Name}} scored {{.Score}}. He did {{if lt .Score 50}}bad{{else if lt
.Score 75}}okay{{else if lt .Score 90}}good{{else}}great{{end}}
```

```
{{end}}`)
if err != nil {
    log.Fatalln("Error:", err)
}
if err = tpl.Execute(os.Stdout, data); err != nil {
    log.Fatalln("Error:", err)
}
```

The full example is available here: `https://play.golang.org/p/eBKDcJ47rPU`.

We will explore functions in more detail in the next section.

Template functions

Functions are an important part of the template engine and there are many built-in functions, such as comparison (`eq`, `lt`, `gt`, `le`, `ge`) or logical (`AND`, `OR`, `NOT`). Functions are called by their names, followed by arguments using space as separator. The function used in the previous example, `lt a b`, means `lt(a,b)`. When the functions are more nested, it's required to wrap functions and arguments in parentheses. For instance, the `not lt a b` statement means that the X function has three arguments, `not(lt, a, b)`. The correct version is `not (lt a b)`, which tells the template that the elements in the parentheses need to be solved first.

When creating a template, custom functions can be assigned to it with the `Funcs` method and can be used in the template. This is very useful, as we can see in this example:

```
var data = struct {
    Name, Surname, Occupation, City string
}{
    "Bojack", "Horseman", "Actor", "Los Angeles",
}
tpl, err := template.New("question-answer").Funcs(template.FuncMap{
    "upper": func(s string) string { return strings.ToUpper(s) },
    "lower": func(s string) string { return strings.ToLower(s) },
}).Parse(`{{.Name}} {{.Surname}} - {{lower .Occupation}} from {{upper
.City}}`)
if err != nil {
    log.Fatalln("Error:", err)
}
if err = tpl.Execute(os.Stdout, data); err != nil {
    log.Fatalln("Error:", err)
}
```

The full example is available here: `https://play.golang.org/p/DdoKEOixDDB`.

The | operator can be used to link the output of a statement to the input of another statement, similar to how it happens in the Unix shell. For instance, the {{"put" | printf "%s%s" "out" | printf "%q"}} statement will produce "output".

RPC servers

Remote Procedure Call (RPC) is a method of calling the execution of an application functionality from another system, using the TCP protocol. Go has native support for RPC servers.

Defining a service

The Go RPC server permits us to register any Go type, along with its methods. This exposes the methods with the RPC protocol and enables us to call them by name from a remote client. Let's create a helper for keeping track of our progress in the book as we are reading:

```go
// Book represents a book entry
type Book struct {
    ISBN string
    Title, Author string
    Year, Pages int
}

// ReadingList keeps tracks of books and pages read
type ReadingList struct {
    Books []Book
    Progress []int
}
```

First, let's define a small helper method called bookIndex that returns the index of a book using its identifier—the ISBN:

```go
func (r *ReadingList) bookIndex(isbn string) int {
    for i := range r.Books {
        if isbn == r.Books[i].ISBN {
            return i
        }
    }
    return -1
}
```

Now, we can define the operation that `ReadingList` will be capable of. It should be able to add and remove books:

```
// AddBook checks if the book is not present and adds it
func (r *ReadingList) AddBook(b Book) error {
    if b.ISBN == "" {
        return ErrISBN
    }
    if r.bookIndex(b.ISBN) != -1 {
        return ErrDuplicate
    }
    r.Books = append(r.Books, b)
    r.Progress = append(r.Progress, 0)
    return nil
}

// RemoveBook removes the book from list and forgets its progress
func (r *ReadingList) RemoveBook(isbn string) error {
    if isbn == "" {
        return ErrISBN
    }
    i := r.bookIndex(isbn)
    if i == -1 {
        return ErrMissing
    }
    // replace the deleted book with the last of the list
    r.Books[i] = r.Books[len(r.Books)-1]
    r.Progress[i] = r.Progress[len(r.Progress)-1]
    // shrink the list of 1 element to remove the duplicate
    r.Books = r.Books[:len(r.Books)-1]
    r.Progress = r.Progress[:len(r.Progress)-1]
    return nil
}
```

It should also be able to read and alter a book's progress:

```
// GetProgress returns the progress of a book
func (r *ReadingList) GetProgress(isbn string) (int, error) {
  if isbn == "" {
  return -1, ErrISBN
  }
  i := r.bookIndex(isbn)
  if i == -1 {
  return -1, ErrMissing
  }
  return r.Progress[i], nil
}
```

Then, `SetProgress` changes the progress of a book, as follows:

```go
func (r *ReadingList) SetProgress(isbn string, pages int) error {
if isbn == "" {
return ErrISBN
}
i := r.bookIndex(isbn)
if i == -1 {
return ErrMissing
}
if p := r.Books[i].Pages; pages > p {
pages = p
}
r.Progress[i] = pages
return nil
}
```

`AdvanceProgress` adds pages to the progress of a book:

```go
func (r *ReadingList) AdvanceProgress(isbn string, pages int) error {
    if isbn == "" {
        return ErrISBN
    }
    i := r.bookIndex(isbn)
    if i == -1 {
        return ErrMissing
    }
    if p := r.Books[i].Pages - r.Progress[i]; p < pages {
        pages = p
    }
    r.Progress[i] += pages
    return nil
}
```

The error variables we are using in these functions are defined as follows:

```go
// List of errors
var (
    ErrISBN = fmt.Errorf("missing ISBN")
    ErrDuplicate = fmt.Errorf("duplicate book")
    ErrMissing = fmt.Errorf("missing book")
)
```

Creating the server

Now we have the service that we can use to create an RPC server very easily. However, the type used has to respect some rules for its methods to make them available:

- The method's type and method itself are both exported.
- The method has two arguments, both exported.
- The second argument is a pointer.
- The method returns an error.

The method should look something like this: `func (t *T) Method(in T1, out *T2) error`.

The next step is create a wrapper for the `ReadingList` that satisfies these rules:

```go
// ReadingService adapts ReadingList for RPC
type ReadingService struct {
    ReadingList
}

// sets the success pointer value from error
func setSuccess(err error, b *bool) error {
    *b = err == nil
    return err
}
```

We can redefine the book, add and remove functions using `Book`, which is an exported type, and built-in types:

```go
func (r *ReadingService) AddBook(b Book, success *bool) error {
    return setSuccess(r.ReadingList.AddBook(b), success)
}

func (r *ReadingService) RemoveBook(isbn string, success *bool) error {
    return setSuccess(r.ReadingList.RemoveBook(isbn), success)
}
```

For the progress, we have two inputs (`isbn` and `pages`), so we have to define a structure that contains both, since the input must be a single argument:

```go
func (r *ReadingService) GetProgress(isbn string, pages *int) (err error) {
    *pages, err = r.ReadingList.GetProgress(isbn)
    return err
}

type Progress struct {
```

```
        ISBN string
        Pages int
}

func (r *ReadingService) SetProgress(p Progress, success *bool) error {
    return setSuccess(r.ReadingList.SetProgress(p.ISBN, p.Pages), success)
}

func (r *ReadingService) AdvanceProgress(p Progress, success *bool) error {
    return setSuccess(r.ReadingList.AdvanceProgress(p.ISBN, p.Pages),
success)
}
```

The defined type can be registered and used in an RPC server, which will use
`rpc.HandleHTTP` to register the HTTP handler for the incoming RPC messages:

```
if len(os.Args) != 2 {
    log.Fatalln("Please specify an address.")
}
if err := rpc.Register(&common.ReadingService{}); err != nil {
    log.Fatalln(err)
}
rpc.HandleHTTP()

l, err := net.Listen("tcp", os.Args[1])
if err != nil {
    log.Fatalln(err)
}
log.Println("Server Started")
if err := http.Serve(l, nil); err != nil {
    log.Fatal(err)
}
```

Creating the client

The client can be created using the RPC package's `rpc.DialHTTP` function, using the same
host-port to obtain a client:

```
if len(os.Args) != 2 {
    log.Fatalln("Please specify an address.")
}
client, err := rpc.DialHTTP("tcp", os.Args[1])
if err != nil {
    log.Fatalln(err)
}
defer client.Close()
```

Then, we define a list of books that we are going to use for our example:

```
const hp = "H.P. Lovecraft"
var books = []common.Book{
    {ISBN: "1540335534", Author: hp, Title: "The Call of Cthulhu", Pages:
36},
    {ISBN: "1980722803", Author: hp, Title: "The Dunwich Horror ", Pages:
53},
    {ISBN: "197620299X", Author: hp, Title: "The Shadow Over Innsmouth",
Pages: 40},
    {ISBN: "1540335534", Author: hp, Title: "The Case of Charles Dexter
Ward", Pages: 176},
}
```

Considering that the format package prints the address of pointers to built-in types, we are going to define a helper function to show the pointer's content:

```
func callClient(client *rpc.Client, method string, in, out interface{}) {
    var r interface{}
    if err := client.Call(method, in, out); err != nil {
        out = err
    }
    switch v := out.(type) {
    case error:
        r = v
    case *int:
        r = *v
    case *bool:
        r = *v
    }
    log.Printf("%s: [%+v] -> %+v", method, in, r)
}
```

The client gets the operation to execute in the form of `type.method`, so we will be using the function like this:

```
callClient(client, "ReadingService.GetProgress", books[0].ISBN, new(int))
callClient(client, "ReadingService.AddBook", books[0], new(bool))
callClient(client, "ReadingService.AddBook", books[0], new(bool))
callClient(client, "ReadingService.GetProgress", books[0].ISBN, new(int))
callClient(client, "ReadingService.AddBook", books[1], new(bool))
callClient(client, "ReadingService.AddBook", books[2], new(bool))
callClient(client, "ReadingService.AddBook", books[3], new(bool))
callClient(client, "ReadingService.SetProgress", common.Progress{
    ISBN: books[3].ISBN,
    Pages: 10,
}, new(bool))
callClient(client, "ReadingService.GetProgress", books[3].ISBN, new(int))
```

```
callClient(client, "ReadingService.AdvanceProgress", common.Progress{
    ISBN: books[3].ISBN,
    Pages: 40,
}, new(bool))
callClient(client, "ReadingService.GetProgress", books[3].ISBN, new(int))
```

This will output each operation with its result.

Summary

In this chapter, we examined how network connections are handled in Go. We started with some network standards. First, we discussed the OSI model, and then TCP/IP.

Then, we checked the network package and learned how to use it to create and manage TCP connections. This included the handling of special commands and how to terminate the connection from the server side. Next, we saw how to do the same with UDP, and we have seen how to implement a custom encoding with checksum control.

Then, we discussed the HTTP protocol, explained how the first version works, and then talked about the differences and improvements of HTTP/2. Then, we learned how to make an HTTP request using Go, followed by how to set up a web server. We explored how to serve existing files, how to associate different actions to different HTTP methods, and how to handle multipart requests and file uploads. We set up an HTTPS server easily, and then we learned what advantages are offered by some third-party libraries for web servers. Finally, we demonstrated how the template engine works in Go, and how to build an RPC client/server easily.

In the next chapter, we are going to cover how to use the main data interchange format such as JSON and XML, which can also be used to create web servers.

Questions

1. What's the advantage of using communication models?
2. What's the difference between a TCP and a UDP connection?
3. Who closes the request body when sending requests?
4. Who closes the body when receiving them in the server?

Data Encoding Using Go **10**

This chapter will show you how to use the more common encoding for exchanging data in an application. Encoding is the process of transforming data, and it can be used when an application has to communicate with another—using the same encoding will allow the two programs to understand each other. This chapter will explain how to handle text-based protocols such as JSON, first of all, and then how to use binary protocols such as `gob`.

The following topics will be covered in this chapter:

- Using text-based encoding such as JSON and XML
- Learning about binary encoding such as `gob` and `protobuf`

Technical requirements

This chapter requires Go to be installed and your favorite editor to be set up. For more information, please refer to `Chapter 3`, *An Overview of Go*.

In order to use the protocol buffer, you will need to install the `protobuf` library. Instructions are available at `https://github.com/golang/protobuf`.

Understanding text-based encoding

The most human-readable data serialization format is the text-based format. In this section, we will analyze some of the most used text-based encoding—CSV, JSON, XML, and YAML.

CSV

Comma-separated values (CSV) is a type of text encoding that stores data in a tabular form. Each line is a table entry and values of a line are separated by a special character, which is usually a comma—hence, the name CSV. Each record of the CSV file must have the same value count and the first record can be used as a header to describe each record field:

```
name,age,country
```

String values can be quoted in order to permit the use of the comma.

Decoding values

Go allows users to create a CSV reader from any `io.Reader`. It is possible to read records one by one using the `Read` method:

```
func main() {
    r := csv.NewReader(strings.NewReader("a,b,c\ne,f,g\n1,2,3"))
    for {
        r, err := r.Read()
        if err != nil {
            log.Fatal(err)
        }
        log.Println(r)
    }
}
```

A full example of the preceding code is available at `https://play.golang.org/p/wZgVzMqAN_K`.

Note that each record is a string slice and the reader is expecting the length of each row to be consistent. If a row has more or fewer entries than the first, this will result in an error. It is also possible to read all records at once using `ReadAll`. The same example from before using such a method will look like this:

```
func main() {
  r := csv.NewReader(strings.NewReader("a,b,c\ne,f,g\n1,2,3"))
  records, err := r.ReadAll()
  if err != nil {
  log.Fatal(err)
  }
  for _, r := range records {
  log.Println(r)
  }
}
```

A full example of the preceding code is available at `https://play.golang.org/p/RJ-wxBB5fs6`.

Encoding values

A CSV writer can be created using any `io.Writer`. The resulting writer will be buffered, so, in order to not lose data, it is necessary to call its method, `Flush`: this will ensure that the buffer gets drained and all content goes to the writer.

The `Write` method receives a string slice and encodes it in CSV format. Let's see how it works in the following example:

```
func main() {
    const million = 1000000
    type Country struct {
        Code, Name string
        Population int
    }
    records := []Country{
        {Code: "IT", Name: "Italy", Population: 60 * million},
        {Code: "ES", Name: "Spain", Population: 46 * million},
        {Code: "JP", Name: "Japan", Population: 126 * million},
        {Code: "US", Name: "United States of America", Population: 327 *
million},
    }
    w := csv.NewWriter(os.Stdout)
    defer w.Flush()
    for _, r := range records {
        if err := w.Write([]string{r.Code, r.Name,
```

```
        strconv.Itoa(r.Population)}); err != nil {
                fmt.Println("error:", err)
                os.Exit(1)
            }
        }
    }
```

A full example of the preceding code is available at `https://play.golang.org/p/` `qwaz3xCJhQT`.

As it happens, for the reader, there is a method for writing more than one record at once. It is known as `WriteAll`, and we can see it in the next example:

```
func main() {
    const million = 1000000
    type Country struct {
        Code, Name string
        Population int
    }
    records := []Country{
        {Code: "IT", Name: "Italy", Population: 60 * million},
        {Code: "ES", Name: "Spain", Population: 46 * million},
        {Code: "JP", Name: "Japan", Population: 126 * million},
        {Code: "US", Name: "United States of America", Population: 327 *
million},
    }
    w := csv.NewWriter(os.Stdout)
    defer w.Flush()
    var ss = make([][]string, 0, len(records))
    for _, r := range records {
        ss = append(ss, []string{r.Code, r.Name,
strconv.Itoa(r.Population)})
    }
    if err := w.WriteAll(ss); err != nil {
        fmt.Println("error:", err)
        os.Exit(1)
    }
}
```

A full example of the preceding code is available at `https://play.golang.org/p/lt_` `GBOLvUfk`.

The main difference between `Write` and `WriteAll` is that the second operation uses more resources and it requires us to convert the records into a slice of string slices before calling it.

Custom options

Both reader and writer have some options that can be changed after creation. Both structures share the `Comma` field, which is the character used for separating fields. Another important field that belongs to the writer only is `FieldsPerRecord`, which is an integer that determines how many fields the reader should expect for each record:

- If greater than `0`, it will be the number of field required.
- If equal to `0` it will set to the number of field of the first record.
- If negative, all checks on the field count will be skipped, allowing for the reading of inconsistent sets of records.

Let's look at a practical example of a reader that is not checking for consistency and uses a space as a separator:

```
func main() {
    r := csv.NewReader(strings.NewReader("a b\ne f g\n1"))
    r.Comma = ' '
    r.FieldsPerRecord = -1
    records, err := r.ReadAll()
    if err != nil {
        log.Fatal(err)
    }
    for _, r := range records {
        log.Println(r)
    }
}
```

A full example of the preceding code is available at `https://play.golang.org/p/KPHXRW5OxXT`.

JSON

JavaScript Object Notation (JSON) is a lightweight, text-based data interchange format. Its nature enables humans to read and write it easily, and its small overhead makes it very suitable for web-based applications.

There are two main types of entities that compose JSON:

- **Collections of name/value pairs**: The name/value is represented as an object, structure, or dictionary in various programming languages.
- **Ordered lists of values**: These are lists of collections or values, which are usually represented as arrays or lists.

The objects are enclosed in braces with each key separated by a colon, and each value is comma separated from the next key/value. Lists are enclosed in brackets and elements are comma separated. These two types can be combined, so a list can also be a value, and objects can be elements in a list. Spaces, newlines, and tabs that are outside names and values are ignored and are used to indent the data and make it easier to read.

Take this sample JSON object:

```
{
    "name: "Randolph",
    "surname": "Carter",
    "job": "writer",
    "year_of_birth": 1873
}
```

It could be compressed into one line, removing the indentation, as this is a good practice when the data length matters—such as in a web server or a database:

```
{"name":"Randolph","surname":"Carter","job":"writer","year_of_birth":1873}
```

In Go, the default types associated to JSON dictionaries and lists are `map[string]interface{}` and `[]interface{}`. These two types (being so generic) are capable of hosting any JSON data structure.

Field tags

A `struct` can also host a specific set of JSON data; all of the exported keys will have names identical to the respective fields. In order to customize the keys, Go enables us to follow field declarations in a struct with a string that should contain metadata about the field.

These tags take the form of colon separated keys/values. The value is a quoted string, which can contain additional information added using commas (such as `job, omitempty`). If there is more than one tag, spaces are used to separate them. Let's see a practical example that uses struct tags:

```
type Character struct {
    Name        string `json:"name" tag:"foo"`
    Surname     string `json:"surname"`
    Job         string `json:"job,omitempty"`
    YearOfBirth int    `json:"year_of_birth,omitempty"`
}
```

This example shows how two different tags can be used for the same field (we have both `json` and `foo`), and it shows how to specify a particular JSON key and introduce the `omitempty` tag that is used for output purposes to avoid marshaling the field if it has a zero value.

Decoder

There two ways of decoding data in JSON—the first is the `json.Unmarshal` function that uses a byte slice as input, and the second is the `json.Decoder` type that uses a generic `io.Reader` to get the encoded content. We will use the latter in our examples because it will enable us to work with structs such as `strings.Reader`. Another advantage of the decoder is the customization that can be done with the following methods:

- `DisallowUnknownFields`: The decode will return an error if a field that is unknown to the receiving data structure is found.
- `UseNumber`: Numbers will be stored as `json.Number` instead of `float64`.

This is a practical example of data decoding using the `json.Decoder` type:

```
r := strings.NewReader(`{
    "name":"Lavinia",
    "surname":"Whateley",
    "year_of_birth":1878
}`)
d := json.NewDecoder(r)
var c Character
if err := d.Decode(&c); err != nil {
    log.Fatalln(err)
}
log.Printf("%+v", c)
```

The full example is available here: `https://play.golang.org/p/a-qt5Mk9E_J`.

Encoder

Data encoding works in a similar fashion, with a `json.Marshal` function that takes a byte slice and the `json.Encoder` type that uses `io.Writer` instead. The latter is better for the obvious reasons of flexibility and customization. It allows us to change the output with the following methods:

- `SetEscapeHTML`: If true, it specifies whether a problematic HTML character should be escaped inside JSON quoted strings.

- `SetIndent`: This allows us to specify a prefix at the beginning of each line, and what string will be used to indent the output JSON.

The following example uses an encore to marshal a data structure to standard output using tabs for indentation:

```
e := json.NewEncoder(os.Stdout)
e.SetIndent("", "\t")
c := Character{
    Name: "Charles Dexter",
    Surname: "Ward",
    YearOfBirth: 1902,
}
if err := e.Encode(c); err != nil {
    log.Fatalln(err)
}
```

This is where we can see the utility of the `omitempty` tag in the `Job` field. Since the value is an empty string, its encoding is skipped. If the tag was absent, there would have been the `"job":""`, line after the surname.

Marshaler and unmarshaler

Encoding and decoding are normally done using the reflection package, which is pretty slow. Before resorting to it, the encoder and decoder will check whether the data type implements the `json.Marshaller` and `json.Unmarshaller` interfaces and use the respective methods instead:

```
type Marshaler interface {
        MarshalJSON() ([]byte, error)
}

type Unmarshaler interface {
        UnmarshalJSON([]byte) error
}
```

Implementing this interface allows for a generally faster encoding and decoding, and it grants the ability to execute other kinds of actions that would not be possible otherwise, such as reading from or writing to unexported fields; it can also embed some operations such as running checks on the data.

If the goal is just to wrap the default behavior, it is necessary to define another type with the same data structure, so that it loses all methods. Otherwise, calling `Marshal` or `Unmarshal` inside the methods will result in a recursive call and, finally, a stack overflow.

In this practical example, we are defining a custom `Unmarshal` method to set a default value for the `Job` field when it's empty:

```
func (c *Character) UnmarshalJSON(b []byte) error {
    type C Character
    var v C
    if err := json.Unmarshal(b, &v); err != nil {
        return err
    }
    *c = Character(v)
    if c.Job == "" {
        c.Job = "unknown"
    }
    return nil
}
```

The full example is available here: `https://play.golang.org/p/4BjFKiMiVHO`.

The `UnmarshalJSON` method needs a pointer receiver because it has to actually modify the value of the data type, but for the `MarshalJSON` method, there's no real need for it, and it is a good practice to have a value receiver—unless the data type should do something different while `nil`:

```
func (c Character) MarshalJSON() ([]byte, error) {
    type C Character
    v := C(c)
    if v.Job == "" {
        v.Job = "unknown"
    }
    return json.Marshal(v)
}
```

The full example is available here: `https://play.golang.org/p/Q-q-9y6v6u-`.

Interfaces

When working with interface types, the encoding part is really straightforward because the application knows which data structure is stored within the interface and will proceed with the marshaling. Doing the opposite operation is not quite as straightforward because the application received has an interface rather than a data structure and does not know what to do and, therefore, ends up doing nothing.

A strategy that works really well (even if it involves a little boilerplate) is using a concrete type container, which will permit us to handle the interface in the `UnmarshalJSON` method. Let's create a quick example by defining an interface and some different implementations:

```
type Fooer interface {
    Foo()
}

type A struct{ Field string }

func (a *A) Foo() {}

type B struct{ Field float64 }

func (b *B) Foo() {}
```

Then, we define a type that wraps the interface and has a `Type` field:

```
type Wrapper struct {
    Type string
    Value Fooer
}
```

Then, let's populate the `Type` field before encoding:

```
func (w Wrapper) MarshalJSON() ([]byte, error) {
    switch w.Value.(type) {
    case *A:
        w.Type = "A"
    case *B:
        w.Type = "B"
    default:
        return nil, fmt.Errorf("invalid type: %T", w.Value)
    }
    type W Wrapper
    return json.Marshal(W(w))
}
```

The decoding method is the more important one: it uses `json.RawMessage`, which is a special type of byte slice used for lazy decoding. We will first get the type from the string field and leave the value in the raw format, in order to use the correct data structure for its decoding:

```
func (w *Wrapper) UnmarshalJSON(b []byte) error {
    var W struct {
        Type string
        Value json.RawMessage
```

```
    }
    if err := json.Unmarshal(b, &W); err != nil {
        return err
    }
    var value interface{}
    switch W.Type {
    case "A":
        value = new(A)
    case "B":
        value = new(B)
    default:
        return fmt.Errorf("invalid type: %s", W.Type)
    }
    if err := json.Unmarshal(W.Value, &value); err != nil {
        return err
    }
    w.Type, w.Value = W.Type, value.(Fooer)
    return nil
}
```

The full example is available here: `https://play.golang.org/p/GXMK_hC8Bpv`.

Generating structs

There is a very useful application that, when given a JSON string, generates a Go type automatically trying to infer field types. You can find one deployed at this address: `https://mholt.github.io/json-to-go/`.

It saves some time and most of the time the data structure is already correct after a simple conversion. Sometimes, it needs some changes such as number types, for instance, if you want a field to be `float`, but your sample JSON has an integer.

JSON schemas

A JSON schema is a vocabulary that describes JSON data and enables the verification of data validity. It can be used for testing and it can be used as documentation. The schema specifies the type of an element and can add additional checks on its value. If the type is an array, it can also specify the type and details of each element. If the type is an object, it describes its fields. Let's see a JSON schema for the `Character` struct that we used in the examples:

```
{
    "type": "object",
    "properties": {
```

```
        "name": { "type": "string" },
        "surname": { "type": "string" },
        "year_of_birth": { "type": "number"},
        "job": { "type": "string" }
    },
    "required": ["name", "surname"]
}
```

We can see that it specifies an object with all of its fields and indicates which fields are mandatory. There are some third-party Go packages that permit us to verify JSON against schema very easily such as `github.com/xeipuuv/gojsonschema`.

XML

Extensible Markup Language (**XML**) is another widely used data encoding format. It's human and machine readable, like JSON, and it was defined by **World Wide Web Consortium** (**W3C**) in 1996. It focuses on simplicity, usability, and generality, and it is actually used as a base for many formats including RSS or XHTML.

Structure

Each XML file starts with a declaration statement that specifies the version and encoding used in the file and whether the file is standalone (schema used is internal). This is a sample XML declaration:

```
<?xml version="1.0" encoding="UTF-8"?>
```

The declaration is followed by a tree of XML elements, which are delimited by tags that have the following form:

- `<tag>`: Opening tag, defines the start of an element
- `</tag>`: Closing tag, defines the end of an element
- `<tag/>`: Self-closing tag, defines an element with no content

Usually, elements are nested so that there are tags inside other tags:

```
<outer>
    <middle>
        <inner1>content</inner1>
        <inner2/>
    </middle>
</outer>
```

Each element can have additional information in the form of attributes, which are space-separated key/value pairs found inside an opening or self-closing tag. The key and value are separated by an equals sign, and the value is delimited by double quotes. The following are examples of elements with attributes:

```
<tag attribute="value" another="something">content</tag>
<selfclosing a="1000" b="-1"/>
```

Document Type Definition

Document Type Definition (DTD) is an XML document that defines the structure and constraints of other XML documents. It can be used to verify the validity of XML if its content is what is expected. An XML can and should specify its own schema, to ease the validation process. The elements of a DTD are as follows:

- **Schema**: This represents the root of the document.
- **Complex type**: It allows an element to have content.
- **Sequence**: This specifies the child elements that must appear in the described sequence.
- **Element**: This represents an XML element.
- **Attribute**: This represents an XML attribute for the parent tag.

This is a sample schema declaration for the `Character` struct we are using in this chapter:

```
<?xml version="1.0" encoding="UTF-8" ?>
<xs:schema xmlns:xs="http://www.w3.org/2001/XMLSchema">
  <xs:element name="character">
    <xs:complexType>
      <xs:sequence>
        <xs:element name="name" type="xs:string" use="required"/>
        <xs:element name="surname" type="xs:string" use="required"/>
        <xs:element name="year_of_birth" type="xs:integer"/>
        <xs:element name="job" type="xs:string"/>
      </xs:sequence>
      <xs:attribute name="id" type="xs:string" use="required"/>
    </xs:complexType>
  </xs:element>
</xs:schema>
```

We can see that it is a schema with an element (character) that is a complex type composed by a sequence of other elements.

Decoding and encoding

As we already saw for JSON, data decoding and encoding can be achieved in two different ways: by providing or returning a byte slice using `xml.Unmarshal` and `xml.Marshal` or by using `io.Reader` or `io.Writer` with the `xml.Decoder` and `xml.Encoder` types.

We can do it by replacing the `json` tag from the `Character` struct with `xml` or by simply adding them:

```
type Character struct {
    Name        string `xml:"name"`
    Surname     string `xml:"surname"`
    Job         string `xml:"job,omitempty"`
    YearOfBirth int    `xml:"year_of_birth,omitempty"`
}
```

Then, we use `xml.Decoder` to unmarshal the data:

```
r := strings.NewReader(`<?xml version="1.0" encoding="UTF-8"?>
<character>
 <name>Herbert</name>
 <surname>West</surname>
 <job>Scientist</job>
</character>
}`)
d := xml.NewDecoder(r)
var c Character
if err := d.Decode(&c); err != nil {
 log.Fatalln(err)
}
log.Printf("%+v", c)
```

The full example is available here: `https://play.golang.org/p/esopq0SMhG_T`.

When encoding, the `xml` package will get the name of the root node from the data type used. If the data structure has a field named `XMLName`, the relative XML `struct` tag will be used for the root node. So, the data structure becomes the following:

```
type Character struct {
    XMLName     struct{} `xml:"character"`
    Name        string   `xml:"name"`
    Surname     string   `xml:"surname"`
    Job         string   `xml:"job,omitempty"`
    YearOfBirth int      `xml:"year_of_birth,omitempty"`
}
```

The encoding operation is also pretty straightforward:

```
e := xml.NewEncoder(os.Stdout)
e.Indent("", "\t")
c := Character{
    Name:        "Henry",
    Surname:     "Wentworth Akeley",
    Job:         "farmer",
    YearOfBirth: 1871,
}
if err := e.Encode(c); err != nil {
    log.Fatalln(err)
}
```

The full example is available here: https://play.golang.org/p/YgZzdPDoaLX.

Field tags

The name of the root tag can be changed using the XMLName field in a data structure. There are some other features of field tags that can be really useful:

- Tags with – are omitted.
- A tag with the attr option becomes an attribute of the parent element.
- A tag with the innerxml option is written verbatim, useful for lazy decoding.
- The omitempty option works the same as it does for JSON; it will not produce a tag for zero values.
- The tag can contain a path in the XML using > as a separator, as a > b > c.
- Anonymous struct fields are treated as if the fields of its value were in the outer struct.

Let's take a look at a practical example that uses some of these features:

```
type Character struct {
    XMLName     struct{} `xml:"character"`
    Name        string   `xml:"name"`
    Surname     string   `xml:"surname"`
    Job         string   `xml:"details>job,omitempty"`
    YearOfBirth int      `xml:"year_of_birth,attr,omitempty"`
    IgnoreMe    string   `xml:"-"`
}
```

This structure produces the following XML:

```
<character year_of_birth="1871">
  <name>Henry</name>
  <surname>Wentworth Akeley</surname>
  <details>
    <job>farmer</job>
  </details>
</character>
```

The full example is available here: `https://play.golang.org/p/6zd19__M0zF`.

Marshaler and unmarshaler

As we have also seen for JSON, the `xml` package offers some interfaces to customize the behavior of a type during encoding and decoding operations—this can avoid the use of reflection or can be used to establish a different behavior. The interfaces that are offered by the package to obtain this behavior are the following:

```
type Marshaler interface {
    MarshalXML(e *Encoder, start StartElement) error
}

type MarshalerAttr interface {
    MarshalXMLAttr(name Name) (Attr, error)
}

type Unmarshaler interface {
        UnmarshalXML(d *Decoder, start StartElement) error
}

type UnmarshalerAttr interface {
        UnmarshalXMLAttr(attr Attr) error
}
```

There are two pairs of functions—one is used when decoding or encoding the type as an element, while the others are used when it's an attribute. Let's see that in action. First, we define a `MarshalXMLAttr` method for a custom type:

```
type Character struct {
    XMLName struct{} `xml:"character"`
    ID ID `xml:"id,attr"`
    Name string `xml:"name"`
    Surname string `xml:"surname"`
    Job string `xml:"job,omitempty"`
    YearOfBirth int `xml:"year_of_birth,omitempty"`
```

```
}

type ID string

func (i ID) MarshalXMLAttr(name xml.Name) (xml.Attr, error) {
    return xml.Attr{
        Name: xml.Name{Local: "codename"},
        Value: strings.ToUpper(string(i)),
    }, nil
}
```

Then, we marshal some data, and we will see that the attribute name is replaced with
`codename`, and its value is uppercase, as specified by the method:

```
e := xml.NewEncoder(os.Stdout)
e.Indent("", "\t")
c := Character{
    ID: "aa",
    Name: "Abdul",
    Surname: "Alhazred",
    Job: "poet",
    YearOfBirth: 700,
}
if err := e.Encode(c); err != nil {
    log.Fatalln(err)
}
```

The full example is available here: `https://play.golang.org/p/XwJrMozQ6RY`.

Generating structs

As seen for JSON, there is a third-party package that can generate Go structures from
encoded files. For XML we have `https://github.com/miku/zek`.

It takes care of any type of XML data including elements with their attributes, spacing
between elements, or comments.

YAML

YAML is a recursive acronym that stands for **YAML Ain't Markup Language** and it's the
name of another widely used encoding format for data. Part of its success is due to it being
easier to write than JSON and XML, its lightweight nature, and its flexibility.

Structure

YAML uses indentation for scope and newlines to separate entities. Elements in a sequence start with a dash followed by a space. The key and value are separated by a color, and a hash sign is used for comments. This is what a sample YAML file can look like:

```
# list of characters
characters:
    - name: "Henry"
      surname: "Armitage"
      year_of_birth: 1855
      job: "librarian"
    - name: "Francis"
      surname: "Wayland Thurston"
      job: "anthropologist"
```

One of the more important differences between JSON and YAML is that, while the first can only use strings as keys, the latter can use any kind of scalar value (strings, numbers, and Booleans).

Decoding and encoding

YAML is not included in the Go standard library but there are many third-party libraries available. The package that is more commonly used to handle this format is the go-yaml package (https://gopkg.in/yaml.v2).

It is built using the following standard encoding packages structure:

- There are encoders and decoders.
- There are Marshal/Unmarshal functions.
- It allows struct tags.
- The behavior of types can be customized by implementing the methods of the interfaces defined.

The interface is slightly different—the Unmarshaler receives the default marshaling function as arguments that can then be used with a data struct that is different to the type:

```
type Marshaler interface {
    MarshalYAML() (interface{}, error)
}

type Unmarshaler interface {
    UnmarshalYAML(unmarshal func(interface{}) error) error
}
```

We can use the `struct` tags in the same way as JSON tags:

```
type Character struct {
    Name        string `yaml:"name"`
    Surname     string `yaml:"surname"`
    Job         string `yaml:"job,omitempty"`
    YearOfBirth int    `yaml:"year_of_birth,omitempty"`
}
```

And we can use them to encode a data structure or, in this case, a list of structures:

```
var chars = []Character{{
    Name:        "William",
    Surname:     "Dyer",
    Job:         "professor",
    YearOfBirth: 1875,
}, {
    Surname: "Danforth",
    Job:     "student",
}}
e := yaml.NewEncoder(os.Stdout)
if err := e.Encode(chars); err != nil {
    log.Fatalln(err)
}
```

Decoding works in the same way, as follows:

```
r := strings.NewReader(`- name: John Raymond
 surname: Legrasse
 job: policeman
- name: "Francis"
 surname: Wayland Thurston
 job: anthropologist`)
// define a new decoder
d := yaml.NewDecoder(r)
var c []Character
// decode the reader
if err := d.Decode(&c); err != nil {
 log.Fatalln(err)
}
log.Printf("%+v", c)
```

We can see that all it takes to create `Decoder` is `io.Reader` and the receiving struct to execute the decode.

Learning about binary encoding

Binary encoding protocols use bytes, so their string representation is not human friendly. They are usually not readable as strings and they are difficult to write, but they are of a smaller size, resulting in faster communication between applications.

BSON

BSON is the binary version of JSON. It is used by MongoDB and has support for some data types that are not available in JSON, such as date and binary.

There are a few packages that implement BSON encoding and decoding, and two of them are very widespread. One is inside the official MongoDB Golang driver, `github.com/mongodb/mongo-go-driver`. The other is not official, but has been around since the beginning of Go and it's part of an unofficial MongoDB driver, `gopkg.in/mgo.v2`.

The second one is very similar to the JSON package in both interfaces and functions. The interfaces are called getter and setter:

- `GetBSON` returns the actual data structure that would be encoded.
- `SetBSON` receives `bson.Raw`, which is a wrapper for `[]byte` that can be used with `bson.Unmarshal`.

A use case for these getters and setters is as follows:

```
type Setter interface {
    SetBSON(raw Raw) error
}

type Getter interface {
    GetBSON() (interface{}, error)
}
```

Encoding

BSON is a format made for documents/entities; therefore, the data structure used for encoding and decoding should be a structure or a map, but not a slice or an array. The mgo version of bson does not offer the usual encoder but only the marshal:

```
var char = Character{
    Name: "Robert",
    Surname: "Olmstead",
}
b, err := bson.Marshal(char)
if err != nil {
    log.Fatalln(err)
}
log.Printf("%q", b)
```

Decoding

The same thing applies to the Unmarshal function:

```
r := []byte(",\x00\x00\x00\x02name\x00\a\x00\x00" +
  "\x00Robert\x00\x02surname\x00\t\x00\x00\x00" +
  "Olmstead\x00\x00")
var c Character
if err := bson.Unmarshal(r, &c); err != nil {
  log.Fatalln(err)
}
log.Printf("%+v", c)
```

gob

gob encoding is another type of binary encoding that is built into the standard library, and it's actually introduced by Go itself. It is a stream of data items, each one preceded by a type declaration, and it does not allow pointers. It uses their value, forbidding the use of nil pointers (since they have no value). The package also has problems with types that have pointers that create a recursive structure and this could lead to unwanted behaviors.

Numbers have arbitrary precision and they can be a float, signed, or unsigned. Signed integers can be stored in any signed integer type, unsigned integers into any unsigned integer type, and floating-point values may be received into any floating-point variable. However, if the variable can't represent the value (overflow, for instance), the decode will fail. Strings and byte slices are stored with a very efficient representation that tries to reuse the same underlying array. Structs will decode only the exported fields, so functions and channels will be ignored.

Interfaces

The interface that gob uses to replace the default marshal and unmarshal behaviors are found in the encoding package:

```
type BinaryMarshaler interface {
        MarshalBinary() (data []byte, err error)
}

type BinaryUnmarshaler interface {
        UnmarshalBinary(data []byte) error
}
```

During the decoding phase, any struct fields that are not present are ignored, since the field name is also part of the serialization.

Encoding

Let's try to encode a structure using gob:

```
var char = Character{
    Name:    "Albert",
    Surname: "Wilmarth",
    Job:     "assistant professor",
}
s := strings.Builder{}
e := gob.NewEncoder(&s)
if err := e.Encode(char); err != nil {
    log.Fatalln(err)
}
log.Printf("%q", s.String())
```

Decoding

Decoding data is also very easy; it works in the same way as the other encoding packages we have already seen:

```
r := strings.NewReader("D\xff\x81\x03\x01\x01\tCharacter" +
    "\x01\xff\x82\x00\x01\x04\x01\x04Name" +
    "\x01\f\x00\x01\aSurname\x01\f\x00\x01\x03" +
    "Job\x01\f\x00\x01\vYearOfBirth\x01\x04\x00" +
    "\x00\x00*\xff\x82\x01\x06Albert\x01\bWilmarth" +
    "\x01\x13assistant professor\x00")
d := gob.NewDecoder(r)
var c Character
if err := d.Decode(&c); err != nil {
    log.Fatalln(err)
}
log.Printf("%+v", c)
```

Now, let's try to decode the same data in different structures—the original one and some with extra or missing fields. We will do this to see how the package behaves. Let's define a general decode function and pass different type of structs to the decoder:

```
func runDecode(data []byte, v interface{}) {
    if err := gob.NewDecoder(bytes.NewReader(data)).Decode(v); err != nil {
        log.Fatalln(err)
    }
    log.Printf("%+v", v)
}
```

Let's try to change the order of the fields in the struct to see whether the gob decoder still works:

```
runDecode(data, new(struct {
    YearOfBirth int      `gob:"year_of_birth,omitempty"`
    Surname     string `gob:"surname"`
    Name        string `gob:"name"`
    Job         string `gob:"job,omitempty"`
}))
```

Let's remove some fields:

```
runDecode(data, new(struct {
    Name string `gob:"name"`
}))
```

Let's put a field in between:

```
runDecode(data, new(struct {
    Name        string `gob:"name"`
    Surname     string `gob:"surname"`
    Country     string `gob:"country"`
    Job         string `gob:"job,omitempty"`
    YearOfBirth int    `gob:"year_of_birth,omitempty"`
}))
```

We can see that the package keeps working even if we scramble, add, or remove the fields. But if we try to change the type of an existing field into another, it fails:

```
runDecode(data, new(struct {
    Name []byte `gob:"name"`
}))
```

Interfaces

Another note about this package is that if you are working with interfaces, their implementation should be registered first, using the functions:

```
func Register(value interface{})
func RegisterName(name string, value interface{})
```

This will make the package aware of the specified types and it will enable us to call decode on the interface type. Let's start by defining an interface and its implementation for our struct:

```
type Greeter interface {
    Greet(w io.Writer)
}

type Character struct {
    Name        string `gob:"name"`
    Surname     string `gob:"surname"`
    Job         string `gob:"job,omitempty"`
    YearOfBirth int    `gob:"year_of_birth,omitempty"`
}

func (c Character) Greet(w io.Writer) {
    fmt.Fprintf(w, "Hello, my name is %s %s", c.Name, c.Surname)
    if c.Job != "" {
        fmt.Fprintf(w, " and I am a %s", c.Job)
    }
}
```

If we try to run the following code without the `gob.Register` function, it returns an error:

```
gob: name not registered for interface: "main.Character"
```

But if we register the type, it works like a charm. Note that the data has been obtained by encoding a pointer to `Greeter` containing the `Character` structure:

```
func main() {
    gob.Register(Greeter(Character{}))
    r := strings.NewReader("U\x10\x00\x0emain.Character" +
        "\xff\x81\x03\x01\x01\tCharacter\x01\xff\x82\x00" +
        "\x01\x04\x01\x04Name\x01\f\x00\x01\aSurname" +
        "\x01\f\x00\x01\x03Job\x01\f\x00\x01\vYearOfBirth" +
        "\x01\x04\x00\x00\x00\x1f\xff\x82\x1c\x01\x05John" +
        " \x01\aKirowan\x01\tprofessor\x00")
    var char Greeter
    if err := gob.NewDecoder(r).Decode(&char); err != nil {
        log.Fatalln(err)
    }
    char.Greet(os.Stdout)
}
```

Proto

A protocol buffer is a serialization protocol made by Google. It is language- and platform-neutral, with little overhead, and it is very efficient. The idea behind it is to define how the data is structured once, then use some tools to generate the source code for the target language of the application.

Structure

The main file that is needed in order to generate the code is the `.proto` file, which uses a specific syntax. We will focus on the latest version of the protocol syntax, `proto3`.

We specify which version of the syntax of the file is to be used in the first line:

```
syntax = "proto3";
```

Definitions from other files can be used, with the `import` statement:

```
import "google/protobuf/any.proto";
```

The rest of the file contains definitions of messages (that are data types) and services. A service is an interface used to define RPC services:

```
message SearchRequest {
   string query = 1;
   int32 page_number = 2;
   int32 result_per_page = 3;
}

service SearchService {
   rpc Search (SearchRequest) returns (SearchResponse);
}
```

Messages are made up by their fields, and services by their methods. Field types are divided between scalars (which includes various integers, signed integers, float, strings, and Booleans) and other messages. Each field has a number associated to it that is its identifier, which should not be changed once selected, so as to maintain compatibility with the older version of the message.

Using the `reserved` keyword allows us to prevent some fields or IDs from being reused, as this is very useful to avoid bugs or problems:

```
message Foo {
   // lock field IDs
   reserved 2, 15, 9 to 11;
   // lock field names
   reserved "foo", "bar";
}
```

Code generation

In order to generate the code from a `.proto` file, you need the `protoc` application and the official proto generation package:

go get -u github.com/golang/protobuf/protoc-gen-go

The installed package comes with the `protoc-gen-go` command; this enables the `protoc` command to use the `--go_out` flag to produce Go source files in the desired folders. Version 1.4 of Go can specify special comments for automatic generation of code with its `go generate` command, and these comments start with `//go:generate` followed by the command, as in the following example:

```
//go:generate protoc -I=$SRC_PATH --go_out=$DST_DIR source.proto
```

It enables us to specify a source path for import lookup, output directory, and a source file. The paths are relative to the package directory where the comment is found and it can be invoked with the `go generate $pkg` command.

Let's start with a simple `.proto` file:

```
syntax = "proto3";

message Character {
    string name = 1;
    string surname = 2;
    string job = 3;
    int32 year_of_birth = 4;
}
```

And let's create a Go source file in the same folder with the comment for generating the code:

```
package gen

//go:generate protoc --go_out=. char.proto
```

Now, we can generate the `go` command that will produce a file with the same name as the `.proto` file and the `.pb.go` extension. The file will contain Go sources for the types and services defined in the `.proto` file:

```
// Code generated by protoc-gen-go. DO NOT EDIT.
// source: char.proto
...
type Character struct {
  Name        string `protobuf:"bytes,1,opt,name=name"`
  Surname     string `protobuf:"bytes,2,opt,name=surname"`
  Job         string `protobuf:"bytes,3,opt,name=job" json:"job,omitempty"`
  YearOfBirth int32
`protobuf:"varint,4,opt,name=year_of_birth,json=yearOfBirth"`
}
```

Encoding

This package permits us to use the `proto.Buffer` type to encode `pb.Message` values. The types created by `protoc` implement the interface defined so the `Character` type can be used directly:

```
var char = gen.Character{
    Name:       "George",
    Surname:    "Gammell Angell",
```

```
        YearOfBirth: 1834,
        Job:            "professor emeritus",
    }
    b := proto.NewBuffer(nil)
    if err := b.EncodeMessage(&char); err != nil {
        log.Fatalln(err)
    }
    log.Printf("%q", b.Bytes())
```

The resulting encoded data has almost no overhead compared to other encoding.

Decoding

The decode operation also needs to be executed using the `proto.Buffer` methods and the generated type:

```
    b := proto.NewBuffer([]byte(
        "/\n\x06George\x12\x0eGammell Angell" +
        "\x1a\x12professor emeritus \xaa\x0e",
    ))
    var char gen.Character
    if err := b.DecodeMessage(&char); err != nil {
        log.Fatalln(err)
    }
    log.Printf("%+v", char)
```

gRPC protocol

Google uses protocol buffer encoding to build a web protocol called **gRPC**. It is a type of remote procedure call that uses HTTP/2 to establish connections and protocol buffers to marshal and unmarshal data.

The first step is generating code related to the server in the target language. This will produce a server interface and a client working implementation. Next, a server implementation needs to be created manually, and finally, the target language will enable the implementation to be used in a gRPC server and then use the client to connect and interact with it.

There are different examples in the `go-grpc` package, including a client/server pair. The client uses the generated code, which only needs a working gRPC connection to the server, and then it can use the methods specified in the service:

```
conn, err := grpc.Dial(address, grpc.WithInsecure())
if err != nil {
    log.Fatalf("did not connect: %v", err)
}
defer conn.Close()
c := pb.NewGreeterClient(conn)

// Contact the server and print out its response
r, err := c.SayHello(ctx, &pb.HelloRequest{Name: name})
```

The full code is available at `grpc/grpc-go/blob/master/examples/helloworld/greeter_client/main.go`.

The server is an implementation of the client interface:

```
// server is used to implement helloworld.GreeterServer.
type server struct{}

// SayHello implements helloworld.GreeterServer
func (s *server) SayHello(ctx context.Context, in *pb.HelloRequest)
(*pb.HelloReply, error) {
    log.Printf("Received: %v", in.Name)
    return &pb.HelloReply{Message: "Hello " + in.Name}, nil
}
```

This interface implementation can be passed to the generated register function, `RegisterGreeterServer`, together with a valid gRPC server, and it can serve incoming connections using a TCP listener:

```
func main() {
    lis, err := net.Listen("tcp", port)
    if err != nil {
        log.Fatalf("failed to listen: %v", err)
    }
    s := grpc.NewServer()
    pb.RegisterGreeterServer(s, &server{})
    if err := s.Serve(lis); err != nil {
        log.Fatalf("failed to serve: %v", err)
    }
}
```

The full code is available at `grpc/grpc-go/blob/master/examples/helloworld/greeter_server/main.go`.

Summary

In this chapter, we explored the encoding methods offered by the Go standard package and third-party libraries. They can be divided into two main categories. The first is the textual-based encoding methods, which are easy to read and write for both human and machines. However, they have more overhead and tend to be much slower than their counterpart, binary-based encoding. Binary-based encoding methods have little overhead but are not human readable.

In text-based encoding, we find JSON, XML, and YAML. The first two are handled by the standard library, the last needs an external dependency. We explored how Go allows us to specify structure tags to change the default encoding and decoding behaviors, and how to use these tags in these operations. Then, we checked and implemented the interfaces that define custom behavior during the marshal and unmarshal operations. There are some third-party tools that allow us to generate the data structures from a JSON file or JSON schemas, which are JSON files used to define the structure of other JSON documents.

XML is another widely used text format that HTML is based on. We checked the XML syntax and composing elements and then we showed a specific type of document called DTD, which is used for defining the content of other XML files. We learned how encoding and decoding works in XML, and the differences to JSON regarding `struct` tags that allow us to define nested XML elements for a type or to store or load a field from an attribute. We concluded with the text-based encoding with the third-party YAML package.

The first binary-based encoding we showed was BSON, a binary version of JSON that is used by MongoDB (which is handled by third-party packages). `gob` is another binary encoding method but it's part of the Go standard library. We learned that both encoding and decoding, together with the involved interfaces, work in the standard package fashion—similar to JSON and XML.

Finally, we looked at the protocol buffer encoding, how to write a `.proto` file and its Go code generation usage and how to use it encode and decode data. We also introduced a practical example of the gRPC encoding, which takes advantage of this encoding to create client/server applications.

In the next chapter, we will start digging into Go's concurrency model, starting with the built-in types—channels and goroutines.

Questions

1. What's the trade-off between text and binary encodings?
2. How does Go behave with data structure by default?
3. How can this behavior be changed?
4. How does a structure field get encoded in an XML attribute?
5. What operation is needed to decode a `gob` interface value?
6. What is the protocol buffer encoding?

Section 4: Deep Dive into Concurrency

This section focuses on one of the Go's most modern features—concurrency. It shows you the tools the language has, looks at sync and channels, and explains how and when to use each tool.

This section consists of the following chapters:

- Chapter 11, *Dealing with Channels and Goroutines*
- Chapter 12, *Sync and Atomic Package*
- Chapter 13, *Coordination Using Context*
- Chapter 14, *Implementing Concurrency Patterns*

11
Dealing with Channels and Goroutines

This chapter will cover concurrent programming with Go, using its basic built-in functionalities, channels, and goroutines. Concurrency describes the capability of executing different parts of an application during the same time period.

Making software concurrent can be a powerful tool in building system applications because some operations can be started even if others have not yet ended.

The following topics will be covered in this chapter:

- Understanding goroutines
- Exploring channels
- Advantage usage

Technical requirements

This chapter requires Go to be installed and your favorite editor to be set up. For more information, refer to Chapter 3, *An Overview of Go*.

Understanding goroutines

Go is a language that centers around concurrency, to the point where two of the main features—channels and goroutines—are part of the built-in package. We will now see how they work and what their basic functionalities are, starting with goroutines, which make it possible to execute parts of an application concurrently.

Comparing threads and goroutines

Goroutines are one of the primitives used for concurrency, but how do they differ from threads? Let's read about each of them here.

Threads

Current OSes are built for modern architectures that have processors with more than one core per CPU, or use technologies, such as hyper-threading, that allow a single core to support more than one thread. Threads are parts of processes that can be managed by the OS scheduler, which can assign them to a specific core/CPU. Like processes, threads carry information about the application execution, but the size of such information is smaller than processes. This includes the current instruction in the program, a stack of the current execution, and the variables needed for it.

The OS is already responsible for the context switch between processes; it saves the older process information and loads the new process information. This is called a **process context switch** and it's a very costly operation, even more than process execution.

In order to jump from one thread to another, the same operation can be done between threads. This is called a **thread context switch** and it is also a heavy operation—even if it is not as hefty as the process switch—because a thread carries less information than a process.

Goroutines

Threads have a minimum size in memory; usually, it is in the order of MBs (2 MB for Linux). The minimum size sets some limitations on the application creation of a new thread—if each thread is at least some MBs, 1,000 threads will occupy at least a few GBs of memory. The way that Go tackles these issues is through the use of a construct similar to threads, but this is handled by the language runtime instead of the OS. The size of a goroutine in memory is three orders of magnitude (2 KB per goroutine), meaning that the minimum memory usage of 1,000 goroutines is comparable to the memory usage of a single thread.

This is obtained by defining what data the goroutines are retaining internally, using a data structure called g that describes the goroutine information, such as stack and status. This is an unexported data type in the runtime package and it can be found in the Go source code. Go keeps a track of OSes using another data structure from the same package called m. The logical processors that are acquired in order to execute a goroutine are stored in p structures. This can be verified in the Go runtime package documentation:

- type g: golang.org/pkg/runtime/?m=all#m
- type m: golang.org/pkg/runtime/?m=all#g
- type p: golang.org/pkg/runtime/?m=all#p

These three entities interact as follows—for each goroutine, a new g gets created, g is queued into p, and each p tries to acquire m to execute the code from g. There are some operations that will block the execution, such as these:

- Built-in synchronization (channels and the sync package)
- System calls that are blocking, such as file operations
- Network operations

When these kinds of operations happen, the runtime detaches p from m and uses (or creates, if it does not already exist) another dedicated m for the blocking operation. The thread becomes idle after executing such operations.

New goroutine

Goroutines are one of the best examples of how Go hides complexity behind a simple interface. When writing an application in order to launch a goroutine, all that is needed is to execute a function preceded by the go keyword:

```
func main() {
    go fmt.Println("Hello, playground")
}
```

The full example is available at https://play.golang.org/p/3gPGZkJtJYv.

If we run the application of the previous example, we will see that it does not produce any output. Why? In Go, the application terminates when the main goroutine does, and it looks like this is the case. What happens is that the Go statements create the goroutine with the respective runtime.g, but this has to be picked up by the Go scheduler, and this does not happen because the program terminates right after the goroutine has been instantiated.

Using the time.Sleep function to let the main goroutine wait (even a nanosecond!) is enough to let the scheduler pick up the goroutine and execute its code. This is shown in the following code:

```go
func main() {
    go fmt.Println("Hello, playground")
    time.Sleep(time.Nanosecond)
}
```

The full example is available at https://play.golang.org/p/2u125pTclv6.

We already saw that Go methods also count as functions, which is why they can be executed concurrently the with go statement, as they were normal functions:

```go
type a struct{}

func (a) Method() { fmt.Println("Hello, playground") }

func main() {
    go a{}.Method()
    time.Sleep(time.Nanosecond)
}
```

The full example is available at https://play.golang.org/p/RUhgfRAPa2b.

Closures are anonymous functions, so they can be used as well, which is actually a very common practice:

```go
func main() {
    go func() {
        fmt.Println("Hello, playground")
    }()
    time.Sleep(time.Nanosecond)
}
```

The full example is available at https://play.golang.org/p/a-JvOVwAwUV.

Multiple goroutines

Organizing code in multiple goroutines can be helpful to split the work between processors and has many other advantages, as we will see in the next chapters. Since they are so lightweight, we can create a number of goroutines very easily using loops:

```
func main() {
    for i := 0; i < 10; i++ {
        go fmt.Println(i)
    }
    time.Sleep(time.Nanosecond)
}
```

The full example is available at `https://play.golang.org/p/Jaljd1padeX`.

This example prints a list of numbers from 0 to 9 in parallel, using concurrent goroutines instead of doing the same thing sequentially in a single goroutine.

Argument evaluation

If we change this example slightly by using a closure without arguments, we will see a very different result:

```
func main() {
    for i := 0; i < 10; i++ {
        go func() { fmt.Println(i) }()
    }
    time.Sleep(time.Nanosecond)
}
```

The full example is available at `https://play.golang.org/p/RV54AsYY-2y`.

If we run this program, we can see that the Go compiler issues a warning in the loop: `loop variable i captured by func literal`.

The variable in the loop gets referenced in the function we defined—the creation loop of the goroutines is quicker than goroutines executing, and the result is that the loop finishes before a single goroutine is started, resulting in the print of the value of the loop variable after the last iteration.

In order to avoid the error of the captured loop variable, it's better to pass the same variable as an argument to the closure. The arguments of the goroutine function are evaluated upon creation, meaning that changes to that variable will not be reflected inside the goroutine, unless you are passing a reference to a value such as a pointer, map, slice, channel, or function. We can see this difference by running the following example:

```
func main() {
    var a int
    // passing value
    go func(v int) { fmt.Println(v) }(a)
    // passing pointer
    go func(v *int) { fmt.Println(*v) }(&a)
    a = 42
    time.Sleep(time.Nanosecond)
}
```

The full example is available at `https://play.golang.org/p/r1dtBiTUMaw`.

Passing the argument by value is not influenced by the last assignment of the program, while passing a pointer type implies that changes done to the pointer content will be seen by the goroutine.

Synchronization

Goroutines allow code to be executed concurrently, but the synchronization between values is not ensured out of the box. We can check out what happens when trying to use a variable concurrently with the following example:

```
func main() {
    var i int
    go func(i *int) {
        for j := 0; j < 20; j++ {
            time.Sleep(time.Millisecond)
            fmt.Println(*i, j)
        }
    }(&i)
    for i = 0; i < 20; i++ {
        time.Sleep(time.Millisecond)
        fmt.Println(i)
    }
}
```

We have an integer variable that changes in the main routine—doing a millisecond pause between each operation—and after the change, the value is printed.

In another goroutine, there is a similar loop (using another variable) and another `print` statement that compares the two values. Considering that the pauses are the same, we would expect to see the same values, but this is not the case. We see that sometimes, the two goroutines are out of sync.

The changes are not reflected immediately because the memory is not synchronized instantaneously. We will learn how to ensure data synchronization in the next chapter.

Exploring channels

Channels are a concept that is unique to Go and a few other programming languages. Channels are very powerful tools that allow a simple method for synchronizing different goroutines, which is one of the ways we can solve the issue raised by the previous example.

Properties and operations

A channel is a built-in type in Go that is typed as arrays, slices, and maps. It is presented in the form of `chan type` and initialized by the `make` function.

Capacity and size

As well as the type that is traveling through the channel, there is another property that the channel has: its `capacity`. This represents the number of items a channel can hold before any new attempt to send an item is made, resulting in a blocking operation. The capacity of the channel is decided at its creation and its default value is `0`:

```
// channel with implicit zero capacity
var a = make(chan int)

// channel with explicit zero capacity
var a = make(chan int, 0)

// channel with explicit capacity
var a = make(chan int, 10)
```

The capacity of the channel cannot be changed after its creation and can be read at any time using the built-in `cap` function:

```
func main() {
    var (
        a = make(chan int, 0)
```

```
        b = make(chan int, 5)
    )

    fmt.Println("a is", cap(a))
    fmt.Println("b is", cap(b))
}
```

The full example is available at `https://play.golang.org/p/Yhz4bTxm5L8`.

The `len` function, when used on a channel, tells us the number of elements that are held by the channel:

```
func main() {
    var (
        a = make(chan int, 5)
    )
    for i := 0; i < 5; i++ {
        a <- i
        fmt.Println("a is", len(a), "/", cap(a))
    }
}
```

The full example is available at `https://play.golang.org/p/zJCL5VGmMsC`.

From the previous example, we can see how the channel capacity remains as 5 and the length grows with each element.

Blocking operations

If a channel is full or its capacity is 0, then the operation will block. If we take the last example, which fills the channel and tries to execute another send operation, our application gets stuck:

```
func main() {
    var (
        a = make(chan int, 5)
    )
    for i := 0; i < 5; i++ {
        a <- i
        fmt.Println("a is", len(a), "/", cap(a))
    }
    a <- 0 // Blocking
}
```

The full example is available at `https://play.golang.org/p/uSfm5zWN8-x`.

When all goroutines are locked (in this specific case, we only have the main goroutine), the Go runtime raises a deadlock—a fatal error that terminates the execution of the application:

```
fatal error: all goroutines are asleep - deadlock!
```

This is can happen with both receive or send operations, and it's the symptom of an error in the application design. Let's take the following example:

```
func main() {
    var a = make(chan int)
    a <- 10
    fmt.Println(<-a)
}
```

In the previous example, there is the a <- 10 send operation and the matching <-a receive operation, but nevertheless, it results in a deadlock. However, the channel we created has no capacity, so the first send operation will block. We can intervene here in two ways:

- **By increasing the capacity**: This is a pretty easy solution that involves initializing the channel with make(chan int, 1). It works best only if the number of receivers is known a priori; if it is higher than the capacity, then the problem appears again.
- **By making the operations concurrent**: This is a far better approach because it uses the channels for what they made for—concurrency.

Let's try to make the previous example work by using the second approach:

```
func main() {
    var a = make(chan int)
    go func() {
        a <- 10
    }()
    fmt.Println(<-a)
}
```

Now, we can see that there are no deadlocks here and the program prints the values correctly. Using the capacity approach will also make it work, but it will be tailored to the fact that we are sending a single message, while the other method will allow us to send any number of messages through the channel and receive them accordingly from the other side:

```
func main() {
    const max = 10
    var a = make(chan int)

    go func() {
        for i := 0; i < max; i++ {
```

```
            a <- i
        }
    }()
    for i := 0; i < max; i++ {
        fmt.Println(<-a)
    }
}
```

The full example is available at https://play.golang.org/p/RKcojupCruB.

We now have a constant to store the number of operations executed, but there is a better and more idiomatic way to let a receiver know when there are no more messages. We will cover this in the next chapter about synchronization.

Closing channels

The best way of handling the end of a synchronization between a sender and a receiver is the close operation. This function is normally executed by the sender because the receiver can verify whether the channel is still open each time it gets a value using a second variable:

```
value, ok := <-ch
```

The second receiver is a Boolean that will be true if the channel is still open, and false otherwise. When a receive operation is done on a close channel, the second received variable will have the false value, and the first one will have the 0 value of the channel type, such as these:

- 0 for numbers
- false for Booleans
- "" for strings
- nil for slices, maps, or pointers

The example of sending multiple messages can be rewritten using the close function, without having prior knowledge of how many messages will be sent:

```
func main() {
    const max = 10
    var a = make(chan int)

    go func() {
        for i := 0; i < max; i++ {
            a <- i
        }
```

```
        close(a)
    }()
    for {
        v, ok := <-a
        if !ok {
            break
        }
        fmt.Println(v)
    }
}
```

The full example is available at `https://play.golang.org/p/GUzgG4kf5ta`.

There is a more synthetic and elegant way to receive a message from a channel until it's closed: by using the same keyword that we used to iterate maps, arrays, and slices. This is done through `range`:

```
for v := range a {
    fmt.Println(v)
}
```

One-way channels

Another possibility when handling channel variables is specifying whether they are only for sending or only for receiving data. This is indicated by the `<-` arrow, which will precede `chan` if it's just for receiving, or follow it if it's just for sending:

```
func main() {
    var a = make(chan int)
    s, r := (chan<- int)(a), (<-chan int)(a)
    fmt.Printf("%T - %T", s, r)
}
```

The full example is available at `https://play.golang.org/p/ZgEPZ99PLJv`.

Channels are already pointers, so casting one of them to its send-only or receive-only version will return the same channel, but will reduce the number of operations that can be performed on it. The types of channels are as follows:

- Send only channels, `chan<-`, which will allow you to send items, close the channel, and prevent you from sending data with a compile error.
- Receive only channel, `<-chan`, that will allow you to receive data, and any send or close operations will be compiling errors.

When a function argument is a send/receive channel, the conversion is implicit and it is a good practice to adopt because it prevents mistakes such as closing the channel from the receiver. We can take the other example and make use of the one-way channels with some refactoring.

We can also create a function for sending values that uses a send-only channel:

```
func send(ch chan<- int, max int) {
    for i := 0; i < max; i++ {
        ch <- i
    }
    close(ch)
}
```

Do the same thing for receiving using a receive-only channel:

```
func receive(ch <-chan int) {
    for v := range ch{
        fmt.Println(v)
    }
}
```

And then, use them with the same channel that will be automatically converted in the one-way version:

```
func main() {
    var a = make(chan int)

    go send(a, 10)
    receive(a)
}
```

The full example is available at https://play.golang.org/p/pPuqpfnq8jJ.

Waiting receiver

Most of the examples we saw in the previous section had the sending operations done in a goroutine, and had the receiving operations done in the main goroutine. It could be the case that all operations are handled by goroutines, so do we synchronize the main one with the others?

A typical technique is the use of another channel used for the sole purpose of signaling that a goroutine has finished its job. The receiving goroutine knows that there are no more messages to get with the closure of the communication channel and it closes another channel that is shared with the main goroutine after finishing its operation. The `main` function can wait for the closure of the channel before exiting.

The typical channel that is used for this scope does not carry any additional information except for whether it is open or closed, so it is usually a `chan struct{}` channel. This is because an empty data structure has no size in memory. We can see this pattern in action by making some changes to the previous example, starting with the receiver function:

```
func receive(ch <-chan int, done chan<- struct{}) {
    for v := range ch {
        fmt.Println(v)
    }
    close(done)
}
```

The receiver function gets an extra argument—the channel. This is used to signal that the sender is done and the `main` function will use that channel to wait for the receiver to end its task:

```
func main() {
    a := make(chan int)
    go send(a, 10)
    done := make(chan struct{})
    go receive(a, done)
    <-done
}
```

The full example is available at `https://play.golang.org/p/thPflJsnKj4`.

Special values

There are a couple of situations in which channels behave differently. We will now see what happens when a channel is set to its zero value—`nil`—or when it is already closed.

nil channels

We have previously discussed how channels belong to the pointer types in Go, so their default value is `nil`. But what happens when you send or receive from a `nil` channel?

If we create a very simple app that tries to send to an empty channel, we get a deadlock:

```
func main() {
    var a chan int
    a <- 1
}
```

The full example is available at `https://play.golang.org/p/KHJ4rvxh7TM`.

If we do the same for a receiving operation, we get the same result of a deadlock:

```
func main() {
    var a chan int
    <-a
}
```

The full example is available at `https://play.golang.org/p/gIjhy7aMxiR`.

The last thing left to check is how the `close` function behaves with a `nil` channel. It panics with the `close of nil channel` explicit value:

```
func main() {
    var a chan int
    close(a)
}
```

The full example is available at `https://play.golang.org/p/5RjdcYUHLSL`.

To recap, we have seen that a `nil` channel's send and receive are blocking operations, and that `close` causes a panic.

Closed channels

We already know that receiving from a closed channel returns a zero value for the channel type, and a second Boolean is `false`. But what happens if we try to send something after closing the channel? Let's find out with the help of the following code:

```
func main() {
    a := make(chan int)
    close(a)
    a <- 1
}
```

The full example is available at `https://play.golang.org/p/_l_xZt1ZojT`.

If we try to send data after closing, it will return a very specific panic: `send on closed channel`. A similar thing will happen when we try to close a channel that has already been closed:

```
func main() {
    a := make(chan int)
    close(a)
    close(a)
}
```

The full example is available at `https://play.golang.org/p/GHK7ERt1XQf`.

This example will cause a panic with a specific value—`close of closed channel`.

Managing multiple operations

There are many situations in which more than one goroutines are executing their code and communicating through channels. A typical scenario is to wait for one of the channels' send or receive operations to be executed.

When you operate with many channels, Go makes it possible to use a special keyword that executes something similar to `switch` but for channel operations. This is done with the `select` statement, followed by a series of `case` statements and an optional `default` case.

We can see a quick example of where we are receiving a value from a channel in a goroutine, and sending a value to another channel in a different goroutine. In these, the main goroutine we are using is a `select` statement to interact with the two channels, receive from the first, and send to the second:

```
func main() {
    ch1, ch2 := make(chan int), make(chan int)
    a, b := 2, 10
    go func() { <-ch1 }()
    go func() { ch2 <- a }()
    select {
    case ch1 <- b:
        fmt.Println("ch1 got a", b)
    case v := <-ch2:
        fmt.Println("ch2 got a", v)
    }
}
```

The full example is available at `https://play.golang.org/p/_8P1Edxe3o4`.

When running this program in the playground, we can see that the receive operation from the second channel is always the first to finish. If we switch the execution order of the goroutines, we get the opposite results. The operation executed last is the one picked up first. This happens because the playground is a web service that runs and executes Go code in a safe environment and does some optimizations to make this operation deterministic.

Default clause

If we add a default case to the previous example, the result of the application execution will be very different, particularly if we change `select`:

```
select {
case v := <-ch2:
    fmt.Println("ch2 got a", v)
case ch1 <- b:
    fmt.Println("ch1 got a", b)
default:
    fmt.Println("too slow")
}
```

The full example is available at `https://play.golang.org/p/F1aE7ImBNFk`.

The `select` statement will always choose the `default` statement. This happens because the goroutines are not picked up by the scheduler yet, when the `select` statement is executed. If we add a very small pause (using `time.Sleep`) before the `select` switch, we will have the scheduler pick at least one goroutine and we will then have one of the two operations executed:

```
func main() {
    ch1, ch2 := make(chan int), make(chan int)
    a, b := 2, 10
    for i := 0; i < 10; i++ {
        go func() { <-ch1 }()
        go func() { ch2 <- a }()
        time.Sleep(time.Nanosecond)
        select {
        case ch1 <- b:
            fmt.Println("ch1 got a", b)
        case v := <-ch2:
            fmt.Println("ch2 got a", v)
        default:
            fmt.Println("too slow")
        }
    }
}
```

The full example is available at https://play.golang.org/p/-aXc3FN6qDj.

In this case, we will have a mixed set of operations executed, depending on which one gets picked up by the Go scheduler.

Timers and tickers

The time package offers a couple of tools that make it possible to orchestrate goroutines and channels—timers and tickers.

Timers

The utility that can replace the default clause in a select statement is the time.Timer type. This contains a receive-only channel that will return a time.Time value after the duration specified during its construction, using time.NewTimer:

```
func main() {
    ch1, ch2 := make(chan int), make(chan int)
    a, b := 2, 10
    go func() { <-ch1 }()
    go func() { ch2 <- a }()
    t := time.NewTimer(time.Nanosecond)
    select {
    case ch1 <- b:
        fmt.Println("ch1 got a", b)
    case v := <-ch2:
        fmt.Println("ch2 got a", v)
    case <-t.C:
        fmt.Println("too slow")
    }
}
```

The full example is available at https://play.golang.org/p/vCAff1kI4yA.

A timer exposes a read-only channel, so it's not possible to close it. When created with time.NewTimer, it waits for the specified duration before firing a value in the channel.

The Timer.Stop method will try to avoid sending data through the channel and return whether it succeeded or not. If false is returned after trying to stop the timer, we still need to receive the value from the channel before being able to use the channel again.

`Timer.Reset` restarts the timer with the given duration, and returns a Boolean as it happens with `Stop`. This value is either `true` or `false`:

- `true` when the timer is active
- `false` when the timer was fired or stopped

We will test these functionalities with a practical example:

```
t := time.NewTimer(time.Millisecond)
time.Sleep(time.Millisecond / 2)
if !t.Stop() {
    panic("it should not fire")
}
select {
case <-t.C:
    panic("not fired")
default:
    fmt.Println("not fired")
}
```

We are creating a new timer of 1ms. Here, we wait 0.5ms and then stop it successfully:

```
if t.Reset(time.Millisecond) {
    panic("timer should not be active")
}
time.Sleep(time.Millisecond)
if t.Stop() {
    panic("it should fire")
}
select {
case <-t.C:
    fmt.Println("fired")
default:
    panic("not fired")
}
```

The full example is available at `https://play.golang.org/p/ddL_fP1UBVv`.

Then, we are resetting the timer back to 1ms and waiting for it to fire, to see whether `Stop` returns `false` and the channel gets drained.

AfterFunc

A very useful utility that uses `time.Timer` is the `time.AfterFunc` function, which returns a timer that will execute the passed function in its own goroutine when the timer fires:

```go
func main() {
    time.AfterFunc(time.Millisecond, func() {
        fmt.Println("Hello 1!")
    })
    t := time.AfterFunc(time.Millisecond*5, func() {
        fmt.Println("Hello 2!")
    })
    if !t.Stop() {
        panic("should not fire")
    }
    time.Sleep(time.Millisecond * 10)
}
```

The full example is available at `https://play.golang.org/p/77HIId1R1Z1`.

In the previous example, we define two timers for two different closures, and we stop one of them and let the other trigger.

Tickers

`time.Ticker` is similar to `time.Timer`, but its channel delivers more elements at regular intervals equal to the duration. They are specified when creating it with `time.NewTicker`. This makes it possible to stop the ticker from firing with the `Ticker.Stop` method:

```go
func main() {
    tick := time.NewTicker(time.Millisecond)
    stop := time.NewTimer(time.Millisecond * 10)
    for {
        select {
        case a := <-tick.C:
            fmt.Println(a)
        case <-stop.C:
            tick.Stop()
        case <-time.After(time.Millisecond):
            return
        }
    }
}
```

The full example is available at `https://play.golang.org/p/8w8I7zIGe-_j`.

In this example, we are also using `time.After`—a function that returns the channel from an anonymous `time.Timer`. This can be used when there's no need to stop the timer. There is another function, `time.Tick`, that returns the channel of an anonymous `time.Ticker`. Both functions will return a channel that the application has no control over and this channel will eventually be picked up by the garbage collector.

This concludes the overview of channels, from their properties and basic usage to some more advanced concurrency examples. We also checked some special cases and how to synchronize multiple channels.

Combining channels and goroutines

Now that we know the fundamental tools and properties of Go concurrency, we can use them to build better tools for our applications. We will see some examples that make use of channels and goroutines to solve real-world problems.

Rate limiter

A typical scenario is having a web API that has a certain limit to the number of calls that can be done in a certain period of time. This type of API will just prevent the usage for a while if this threshold is crossed, making it unusable for the time being. When creating a client for the API, we need to be aware of this and make sure our application does not overuse it.

That's a very good scenario where we can use `time.Ticker` to define an interval between calls. In this example, we will create a client for Google Maps' geocoding service that has a limit of 100,000 requests per 24 hours. Let's start by defining the client:

```
type Client struct {
    client *http.Client
    tick *time.Ticker
}
```

The client is made by an HTTP client that will call maps, a ticker that will help prevent passing the rate limit, and needs an API key for authentication with the service. We can define a custom `Transport` struct for our use case that will inject the key in the request as follows:

```
type apiTransport struct {
    http.RoundTripper
    key string
}

func (a apiTransport) RoundTrip(r *http.Request) (*http.Response, error) {
    q := r.URL.Query()
    q.Set("key", a.key)
    r.URL.RawQuery = q.Encode()
    return a.RoundTripper.RoundTrip(r)
}
```

This is a very good example of how Go interfaces allow the extension of their own behavior. We are defining a type that implements the `http.RoundTripper` interface, and also an attribute that is an instance of the same interface. The implementation injects the API key to the request before executing the underlying transport. This type allows us to define a helper function that creates a new client, where we are using the new transport that we defined together with the default one:

```
func NewClient(tick time.Duration, key string) *Client {
    return &Client{
        client: &http.Client{
            Transport: apiTransport{http.DefaultTransport, key},
        },
        tick: time.NewTicker(tick),
    }
}
```

The maps geocoding API returns a series of addresses that are composed of various parts. This is available at https://developers.google.com/maps/documentation/geocoding/intro#GeocodingResponses.

The result is encoded in JSON, so we need a data structure that can receive it:

```
type Result struct {
    AddressComponents []struct {
        LongName string `json:"long_name"`
        ShortName string `json:"short_name"`
        Types []string `json:"types"`
    } `json:"address_components"`
    FormattedAddress string `json:"formatted_address"`
    Geometry struct {
```

```
        Location struct {
            Lat float64 `json:"lat"`
            Lng float64 `json:"lng"`
        } `json:"location"`
        // more fields
    } `json:"geometry"`
    PlaceID string `json:"place_id"`
    // more fields
}
```

We can use the structure to execute a reverse geocoding operation—getting a location from the coordinates by using the respective endpoint. We wait for the ticket before executing the HTTP request, remembering to `defer` the closure of the body:

```
    const url =
"https://maps.googleapis.com/maps/api/geocode/json?latlng=%v,%v"
    <-c.tick.C
    resp, err := c.client.Get(fmt.Sprintf(url, lat, lng))
    if err != nil {
        return nil, err
    }
    defer resp.Body.Close()
```

Then, we can decode the result in a data structure that uses the `Result` type we already defined and checks for the `status` string:

```
    var v struct {
        Results []Result `json:"results"`
        Status string `json:"status"`
    }
    // get the result
    if err := json.NewDecoder(resp.Body).Decode(&v); err != nil {
        return nil, err
    }
    switch v.Status {
    case "OK":
        return v.Results, nil
    case "ZERO_RESULTS":
        return nil, nil
    default:
        return nil, fmt.Errorf("status: %q", v.Status)
    }
}
```

Finally, we can use the client to geocode a series of coordinates, expecting the requests to be at least 860ms from each other:

```
c := NewClient(24*time.Hour/100000, os.Getenv("MAPS_APIKEY"))
start := time.Now()
for _, l := range [][2]float64{
    {40.4216448, -3.6904040},
    {40.4163111, -3.7047328},
    {40.4123388, -3.7096724},
    {40.4145150, -3.7064412},
} {
    locs, err := c.ReverseGeocode(l[0], l[1])
    e := time.Since(start)
    if err != nil {
        log.Println(e, l, err)
        continue
    }
    // just print the first location
    if len(locs) != 0 {
        locs = locs[:1]
    }
    log.Println(e, l, locs)
}
```

Workers

The previous example is a Google Maps client that uses a `time.Ticker` channel to limit the rate of the requests. The rate limit makes sense for an API key. Let's imagine that we have more API keys from different accounts, so we could potentially execute more requests.

A very typical concurrent approach is the workers pool. Here, you have a series of clients that can be picked up to process an input and different parts of the application can ask to use such clients, returning the clients back when they are done.

We can create more than one client that shares the same channels for both requests and responses, with requests being the coordinates and the results being the response from the service. Since the channel for responses is unique, we can define a custom type that holds all the information needed for that channel:

```
type result struct {
    Loc [2]float64
    Result []maps.Result
    Error error
}
```

The next step is creating the channels—we are going to read a comma-separated list of values from an environment variable here. We will create a channel for requests, and one for responses. Both channels have a capacity equal to the number of workers, in this case, but this would work even if the channels were unbuffered. Since we are just using channels, we will need another channel, done, which signals whether a worker has finished working on their last job:

```
keys := strings.Split(os.Getenv("MAPS_APIKEYS"), ",")
requests := make(chan [2]float64, len(keys))
results := make(chan result, len(keys))
done := make(chan struct{})
```

Now, we will create a goroutine for each of the keys, in which we define a client that feeds on the requests channel, executes the request, and sends the result to the dedicated channel. When the requests channel is closed, the goroutine will exit the range and send a message to the done channel, which is shown in the following code:

```
for i := range keys {
    go func(id int) {
        log.Printf("Starting worker %d with API key %q", id, keys[id])
        client := maps.NewClient(maps.DailyCap, keys[id])
        for j := range requests {
            var r = result{Loc: j}
            log.Printf("w[%d] working on %v", id, j)
            r.Result, r.Error = client.ReverseGeocode(j[0], j[1])
            results <- r
        }
        done <- struct{}{}
    }(i)
}
```

The locations can be sent to the request channel sequentially in another goroutine:

```
go func() {
    for _, l := range [][2]float64{
        {40.4216448, -3.6904040},
        {40.4163111, -3.7047328},
        {40.4123388, -3.7096724},
        {40.4145150, -3.7064412},
    } {
        requests <- l
    }
    close(requests)
}()
```

We can keep count of the done signals we are receiving and close the results channel when all the workers are done:

```
go func() {
    count := 0
    for range done {
        if count++; count == len(keys) {
            break
        }
    }
    close(results)
}()
```

The channel is used to count how many workers are done, and once every one of them is done, it will close the result channel. This will allow us to just loop over it to get the result:

```
for r := range results {
    log.Printf("received %v", r)
}
```

Using a channel is just one of the ways to wait for all the goroutines to finish, and we will see more idiomatic ways of doing it in the next chapter with the `sync` package.

Pool of workers

A channel can be used as a pool of resources that allows us to request them on demand. In the following example, we will create a small application that will look up which addresses are valid in a network, using a third-party client from the `github.com/tatsushid/go-fastping` package.

The pool will have two methods, one for getting a new client and another to return the client back to the pool. The `Get` method will try to get an existing client from the channel or return a new one if this is not available. The `Put` method will try to put the client back in the channel, or discard it otherwise:

```
const wait = time.Millisecond * 250

type pingPool chan *fastping.Pinger

func (p pingPool) Get() *fastping.Pinger {
    select {
    case v := <-p:
        return v
    case <-time.After(wait):
        return fastping.NewPinger()
```

```
    }
}

func (p pingPool) Put(v *fastping.Pinger) {
    select {
    case p <- v:
    case <-time.After(wait):
    }
    return
}
```

The client will need to specify which network needs to be scanned, so it requires a list of available networks starting with the net.Interfaces function, ranging through the interfaces and their addresses:

```
ifaces, err := net.Interfaces()
if err != nil {
    return nil, err
}
for _, iface := range ifaces {
    // ...
    addrs, err := iface.Addrs()
    // ...
    for _, addr := range addrs {
        var ip net.IP
        switch v := addr.(type) {
        case *net.IPNet:
            ip = v.IP
        case *net.IPAddr:
            ip = v.IP
        }
        // ...
        if ip = ip.To4(); ip != nil {
            result = append(result, ip)
        }
    }
}
```

We can accept a command-line argument to select between interfaces, and we can show a list of interfaces to the user to select when the argument is either not present or wrong:

```
if len(os.Args) != 2 {
    help(ifaces)
}
i, err := strconv.Atoi(os.Args[1])
if err != nil {
    log.Fatalln(err)
}
```

```
if i < 0 || i > len(ifaces) {
    help(ifaces)
}
```

The `help` function is just a print of the interfaces IP:

```
func help(ifaces []net.IP) {
    log.Println("please specify a valid network interface number")
    for i, f := range ifaces {
        mask, _ := f.DefaultMask().Size()
        fmt.Printf("%d - %s/%v\n", i, f, mask)
    }
    os.Exit(0)
}
```

The next step is obtain the range of IPs that need to be checked:

```
m := ifaces[i].DefaultMask()
ip := ifaces[i].Mask(m)
log.Printf("Lookup in %s", ip)
```

Now that we have the IP, we can create a function to obtain other IPs in the same network. IPs in Go are a byte slice, so we will replace the least significant bits in order to obtain the final address. Since the IP is a slice, its value will be overwritten by each operation (slices are pointers). We are going to update a copy of the original IP—because slices are pointers to the same array—in order to avoid overwrites:

```
func makeIP(ip net.IP, i int) net.IP {
    addr := make(net.IP, len(ip))
    copy(addr, ip)
    b := new(big.Int)
    b.SetInt64(int64(i))
    v := b.Bytes()
    copy(addr[len(addr)-len(v):], v)
    return addr
}
```

Then, we will need one channel for results and another for keeping a track of the goroutines; and for each IP, we need to check whether we can launch a goroutine for each address. We will use a pool of 10 clients and inside each goroutine—we will ask for each client, then return them to the pool. All valid IPs will be sent through the result channel:

```
done := make(chan struct{})
address := make(chan net.IP)
ones, bits := m.Size()
pool := make(pingPool, 10)
for i := 0; i < 1<<(uint(bits-ones)); i++ {
```

```
go func(i int) {
    p := pool.Get()
    defer func() {
        pool.Put(p)
        done <- struct{}{}
    }()
    p.AddIPAddr(&net.IPAddr{IP: makeIP(ip, i)})
    p.OnRecv = func(a *net.IPAddr, _ time.Duration) { address <- a.IP }
    p.Run()
}(i)
}
```

Each time a routine finishes, we send a value in the done channel so we can keep count of the done signals received before exiting the application. This will be the result loop:

```
i = 0
for {
    select {
    case ip := <-address:
        log.Printf("Found %s", ip)
    case <-done:
        if i >= bits-ones {
            return
        }
        i++
    }
}
```

The loop will continue until the count from the channel reaches the number of goroutines. This concludes the more convoluted examples of the usage of channels and goroutines together.

Semaphores

Semaphores are tools used to solve concurrency issues. They have a certain number of available quotas that is used to limit the access to resources; also, various threads can request one or more quotas from it, and then release them when they are done. If the number of quotas available is one, it means that the semaphore supports only one access at time, with a behavior similar to mutexes. If the quota is more than one, we are referring to the most common type—the weighted semaphore.

In Go, a semaphore can be implemented using a channel with a capacity equal to the quotas, where you send a message to the channel to acquire a quota, and receive one from it to release:

```
type sem chan struct{}

func (s sem) Acquire() {
    s <- struct{}{}
}

func (s sem) Relase() {
    <-s
}
```

The preceding code shows us how to implement a semaphore using a channel in a few lines. Here's an example of how to use it:

```
func main() {
    s := make(sem, 5)
    for i := 0; i < 10; i++ {
        go func(i int) {
            s.Acquire()
            fmt.Println(i, "start")
            time.Sleep(time.Second)
            fmt.Println(i, "end")
            s.Relase()
        }(i)
    }
    time.Sleep(time.Second * 3)
}
```

The full example is available at `https://play.golang.org/p/BR5GN2QopjQ`.

We can see from the previous example how the program serves some requests on the first round of acquisition, and the others on the second round, not allowing more than five executions at the same time.

Summary

In this chapter, we talked about the two main actors in Go concurrency—goroutines and channels. We started by explaining what a thread is, what the differences are between threads and goroutines, and why they are so convenient. Threads are heavy and require a CPU core, while goroutines are lightweight and not bound to a core. We saw how easily a new goroutine can be started by executing a function preceded by the `go` keyword, and how it is possible to start a series of different goroutines at once. We saw how the arguments of the concurrent functions are evaluated when the goroutine is created and not when it actually starts. We also saw that it is very difficult to keep different goroutines in sync without any additional tools.

Then, we introduced channels that are used to share information between different goroutines and solve the synchronization problem that we mentioned previously. We saw that goroutines have a maximum capacity and a size—how many elements it is holding at present. Size cannot overcome capacity, and when an extra element is sent to a full channel, the operation blocks it until an element is removed from the channel. Receiving from a channel that is empty is also a blocking operation.

We saw how to close channels with the `close` function, how this operation should be done in the same goroutine that sends data, and how operations behave in special cases such as `nil` or a closed channel. We introduced the `select` statement to choose between concurrent channel operations and control the application flow. Then, we introduced the tools related to concurrency from the `time` package—tickers and timers.

Finally, we showed some real-world examples, including a rate-limited Google Maps client and a tool to simultaneously ping all the addresses of a network.

In the next chapter, we will look at some synchronization primitives that will allow a better handling of goroutines and memory, using more clear and simple code.

Questions

1. What is a thread and who is responsible for it?
2. Why are goroutines different from threads?
3. When are arguments evaluated when launching a goroutine?
4. What's the difference between buffered and unbuffered channels?
5. Why are one-way channels useful?
6. What happens when operations are done on `nil` or closed channels?
7. What are timers and tickers used for?

Synchronization with sync and atomic 12

This chapter will continue the journey into Go concurrency, introducing the `sync` and `atomic` packages, which are a couple of other tools designed for orchestrating synchronization between goroutines. This will make it possible to write elegant and simple code that allows concurrent usage of resources and manages a goroutine's lifetime. `sync` contains high-level synchronization primitives, while `atomic` contains low-level ones.

The following topics will be covered in this chapter:

- Lockers
- Wait groups
- Other sync components
- The `atomic` package

Technical requirements

This chapter requires Go to be installed and your favorite editor to be set up. For more information on this, refer to `Chapter 3`, *An Overview of Go*.

Synchronization primitives

We saw how channels are focused on communication between goroutines, and now we will focus on the tools offered by the `sync` package, which includes the basic primitives for synchronization between goroutines. The first thing we will see is how to implement concurrent access to the same resource with lockers.

Concurrent access and lockers

Go offers a generic interface for objects that can be locked and unlocked. Locking an object means taking control over it while unlocking releases it for others to use. This interface exposes a method for each operation. The following is an example of this in code:

```
type Locker interface {
    Lock()
    Unlock()
}
```

Mutex

The most simple implementation of locker is `sync.Mutex`. Since its method has a pointer receiver, it should not be copied or passed around by value. The `Lock()` method takes control of the mutex if possible, or blocks the goroutine until the mutex becomes available. The `Unlock()` method releases the mutex and it returns a runtime error if called on a non-locked one.

Here is a simple example in which we launch a bunch of goroutines using the lock to see which is executed first:

```
func main() {
    var m sync.Mutex
    done := make(chan struct{}, 10)
    for i := 0; i < cap(done); i++ {
        go func(i int, l sync.Locker) {
            l.Lock()
            defer l.Unlock()
            fmt.Println(i)
            time.Sleep(time.Millisecond * 10)
            done <- struct{}{}
        }(i, &m)
    }
    for i := 0; i < cap(done); i++ {
        <-done
    }
}
```

The full example is available at: `https://play.golang.org/p/resVh7LImLf`

We are using a channel to signal the main goroutine when a job is done, and exit the application. Let's create an external counter and increment it concurrently using goroutines.

Operations executed on different goroutines are not thread-safe, as we can see from the following example:

```
done := make(chan struct{}, 10000)
var a = 0
for i := 0; i < cap(done); i++ {
    go func(i int) {
        if i%2 == 0 {
            a++
        } else {
            a--
        }
        done <- struct{}{}
    }(i)
}
for i := 0; i < cap(done); i++ {
    <-done
}
fmt.Println(a)
```

We would expect to have 5000 plus one, and 5000 minus one, with a 0 printed in the final instruction. However, what we get are different values each time we run the application. This happens because these kind of operations are not thread-safe, so two or more of them could happen at the same time, with the last one shadowing the others. This kind of phenomena is known as a **race condition**; that is, when more than one operation is trying to write the same result.

This means that without any synchronization, the result is not predictable; if we check the previous example and use a lock to avoid the race condition, we will have zero as the value for the integer—the result that we were expecting:

```
m := sync.Mutex{}
for i := 0; i < cap(done); i++ {
    go func(l sync.Locker, i int) {
        l.Lock()
        defer l.Unlock()
        if i%2 == 0 {
            a++
        } else {
            a--
        }
        done <- struct{}{}
    }(&m, i)
    fmt.Println(a)
}
```

A very common practice is embedding a mutex in a data structure to symbolize that the container is the one you want to lock. The counter variable from before can be represented as follows:

```
type counter struct {
    m       sync.Mutex
    value int
}
```

The operations that the counter performs can be methods that already take care of locking before the main operation, along with unlocking it afterward, as shown in the following code block:

```
func (c *counter) Incr(){
    c.m.Lock()
    c.value++
    c.m.Unlock()
}

func (c *counter) Decr(){
    c.m.Lock()
    c.value--
    c.m.Unlock()
}

func (c *counter) Value() int {
    c.m.Lock()
    a := c.value
    c.m.Unlock()
    return a
}
```

This will simplify the goroutine loop, resulting in a much clearer code:

```
var a = counter{}
for i := 0; i < cap(done); i++ {
    go func(i int) {
        if i%2 == 0 {
            a.Incr()
        } else {
            a.Decr()
        }
        done <- struct{}{}
    }(i)
}
// ...
fmt.Println(a.Value())
```

RWMutex

The problem with race conditions is caused by concurred writing, not by reading the operation. The other data structure that implements the locker interface, `sync.RWMutex`, is made to support both these operations, having write locks that are unique and mutually exclusive with read locks. This means that the mutex can be locked either by a single write lock, or by one or more read locks. When a reader locks the mutex, other readers trying to lock it will not be blocked. They are often referred to as shared-exclusive locks. This allows read operations to happen all at the same time, without there being a waiting time.

The write lock operations are done using the `Lock` and `Unlock` methods of the locker interface. The reading operations are executed using two other methods: `RLock` and `RUnlock`. There is another method, `RLocker`, which returns a locker for reading operations.

We can make a quick example of their usage by creating a concurrent list of strings:

```
type list struct {
    m sync.RWMutex
    value []string
}
```

We can iterate the slice to find the selected value and use a read lock to delay the writing while we are reading:

```
func (l *list) contains(v string) bool {
    for _, s := range l.value {
        if s == v {
            return true
        }
    }
    return false
}

func (l *list) Contains(v string) bool {
    l.m.RLock()
    found := l.contains(v)
    l.m.RUnlock()
    return found
}
```

We can use the write lock when adding new elements:

```
func (l *list) Add(v string) bool {
    l.m.Lock()
    defer l.m.Unlock()
```

```
        if l.contains(v) {
            return false
        }
        l.value = append(l.value, v)
        return true
    }
```

Then we can try to use several goroutines to execute the same operation on the list:

```
var src = []string{
    "Ryu", "Ken", "E. Honda", "Guile",
    "Chun-Li", "Blanka", "Zangief", "Dhalsim",
}
var l list
for i := 0; i < 10; i++ {
    go func(i int) {
        for _, s := range src {
            go func(s string) {
                if !l.Contains(s) {
                    if l.Add(s) {
                        fmt.Println(i, "add", s)
                    } else {
                        fmt.Println(i, "too slow", s)
                    }
                }
            }(s)
        }
    }(i)
}
time.Sleep(500 * time.Millisecond)
```

We are checking whether the name is contained in the lock first, then we try to add the element. This causes more than one routine to attempt to add a new element, but since writing locks are exclusive, only one will succeed.

Write starvation

When designing an application, this kind of mutex is not always the obvious choice, because in a scenario where there is a greater number of read locks and a few write ones, the mutex will be accepting incoming more read locks after the first, letting the write operation wait for a moment where there are no read locks active. This is a phenomenon referred to as **write starvation**.

To check this out, we can define a type that has both a write and a read operation, which take some time, as shown in the following code:

```
type counter struct {
    m sync.RWMutex
    value int
}

func (c *counter) Write(i int) {
    c.m.Lock()
    time.Sleep(time.Millisecond * 100)
    c.value = i
    c.m.Unlock()
}

func (c *counter) Value() int {
    c.m.RLock()
    time.Sleep(time.Millisecond * 100)
    a := c.value
    c.m.RUnlock()
    return a
}
```

We can try to execute both write and read operations with the same cadence in separate goroutines, using a duration that is lower than the execution time of the methods (50 ms versus 100 ms). We will also check out how much time they spend in a locked state:

```
var c counter
t1 := time.NewTicker(time.Millisecond * 50)
time.AfterFunc(time.Second*2, t1.Stop)
for {
    select {
    case <-t1.C:
        go func() {
            t := time.Now()
            c.Value()
            fmt.Println("val", time.Since(t))
        }()
        go func() {
            t := time.Now()
            c.Write(0)
            fmt.Println("inc", time.Since(t))
        }()
    case <-time.After(time.Millisecond * 200):
        return
    }
}
```

If we execute the application, we see that for each write operation, more than one read is executed, and each next call is spending more time than the previous, waiting for the lock. This is not true for the read operation, which can happen at the same time, so as soon as a reader manages to lock the resource, all the other waiting readers will do the same. Replacing `RWMutex` with `Mutex` will make both operations have the same priority, as in the previous example.

Locking gotchas

Some care must be taken when locking and unlocking mutexes in order to avoid unexpected behavior and deadlocks in the application. Take the following snippet:

```
for condition {
    mu.Lock()
    defer mu.Unlock()
    action()
}
```

This code seems okay at first sight, but it will inevitably block the goroutine. This is because the `defer` statement is not executed at the end of each loop iteration, but when the function returns. So the first attempt will lock without releasing and the second attempt will remain stuck.

A little refactor can help fix this, as shown in the following snippet:

```
for condition {
    func() {
        mu.Lock()
        defer mu.Unlock()
        action()
    }()
}
```

We can use a closure to be sure that the deferred `Unlock` gets executed, even if `action` panics.

If the kind of operations that are executed on the mutex will not cause panic, it can be a good idea to ditch the defer and just use it after executing the action, as follows:

```
for condition {
    mu.Lock()
    action()
    mu.Unlock()
}
```

`defer` has a cost, so it is better to avoid it when it is not necessary, such as when doing a simple variable read or assignment.

Synchronizing goroutines

Until now, in order to wait for goroutines to finish, we used a channel of empty structures and sent a value through the channel as the last operation, as follows:

```
ch := make(chan struct{})
for i := 0; i < n; n++ {
    go func() {
        // do something
        ch <- struct{}{}
    }()
}
for i := 0; i < n; n++ {
    <-ch
}
```

This strategy works, but it's not the preferred way to accomplish the task. It's not correct semantically, because we are using a channel, which is a tool for communication, to send empty data. This use case is about synchronization rather than communication. That's why there is the `sync.WaitGroup` data structure, which covers such cases. It has a main status, called counter, which represents the number of elements waiting:

```
type WaitGroup struct {
    noCopy noCopy
    state1 [3]uint32
}
```

The `noCopy` field prevents the structure from being copied by value with `panic`. The state is an array made by three `int32`, but only the first and last entries are used; the remaining one is used for compiler optimizations.

The `WaitGroup` offers three methods to accomplish the same result:

- `Add`: This changes the value of the counter using the given value, which could also be negative. If the counter goes under zero, the application panics.
- `Done`: This is a shorthand for `Add` with `-1` as the argument. It is usually called when a goroutine finishes its job to decrement the counter by 1.
- `Wait`: This operation blocks the current goroutine until the counter reaches zero.

Using the wait group results in a much cleaner and more readable code, as we can see in the following example:

```
func main() {
    wg := sync.WaitGroup{}
    wg.Add(10)
    for i := 1; i <= 10; i++ {
        go func(a int) {
            for i := 1; i <= 10; i++ {
                fmt.Printf("%dx%d=%d\n", a, i, a*i)
            }
            wg.Done()
        }(i)
    }
    wg.Wait()
}
```

To the wait group, we are adding a delta equal to goroutines, which we will launch beforehand. In each single goroutine, we are using the Done method to reduce the count. If the number of goroutines is not known, the Add operation (with 1 as its argument) can be executed before starting each goroutine, as shown in the following:

```
func main() {
    wg := sync.WaitGroup{}
    for i := 1; rand.Intn(10) != 0; i++ {
        wg.Add(1)
        go func(a int) {
            for i := 1; i <= 10; i++ {
                fmt.Printf("%dx%d=%d\n", a, i, a*i)
            }
            wg.Done()
        }(i)
    }
    wg.Wait()
}
```

In the preceding example, we have a 10% chance of finishing each iteration of the for loop, so we are adding one to the group before starting the goroutine.

A very common error is to add the value inside the goroutine, which usually results in a premature exit without any goroutines executed. This happens because the application creates the goroutines and executes the Wait function before the routines start and add their own delta, as in the following example:

```
func main() {
    wg := sync.WaitGroup{}
    for i := 1; i < 10; i++ {
```

```
    go func(a int) {
        wg.Add(1)
        for i := 1; i <= 10; i++ {
            fmt.Printf("%dx%d=%d\n", a, i, a*i)
        }
        wg.Done()
    }(i)
}
wg.Wait()
}
```

This application will not print anything because it arrives at the `Wait` statement before any goroutine is started and the `Add` method is called.

Singleton in Go

The singleton pattern is a commonly used strategy for software development. This involves restricting the number of instances of a certain type to one, using the same instance across the whole application. A very simple implementation of the concept could be the following code:

```
type obj struct {}

var instance *obj

func Get() *obj{
    if instance == nil {
        instance = &obj{}
    }
    return instance
}
```

This is perfectly fine in a consecutive scenario but in a concurrent one, like in many Go applications, this is not thread-safe and could generate race conditions.

The previous example could be made thread-safe by adding a lock that would avoid any race condition, as follows:

```
type obj struct {}

var (
    instance *obj
    lock      sync.Mutex
)

func Get() *obj{
```

```
        lock.Lock()
        defer lock.Unlock()
        if instance == nil {
            instance = &obj{}
        }
        return instance
    }
```

This is safe, but slower, because `Mutex` will be synchronizing each time the instance is requested.

The best solution to implement this pattern, as shown in the following example, is to use the `sync.Once` struct that takes care of executing a function once using a combination of `Mutex` and `atomic` readings (which we will see in the second part of the chapter):

```
type obj struct {}

var (
    instance *obj
    once     sync.Once
)

func Get() *obj{
    once.Do(func(){
        instance = &obj{}
    })
    return instance
}
```

The resulting code is idiomatic and clear, and has better performance compared to the mutex solution. Since the operation will be executed just the once, we can also get rid of the `nil` check we were doing on the instance in the previous examples.

Once and Reset

The `sync.Once` function is made for executing another function once and no more. There is a very useful third-party library, which allows us to reset the state of the singleton using the `Reset` method.

The package source code can be found at: `github.com/matryer/resync`.

Typical uses include some initialization that needs to be done again on a particular error, such as obtaining an API key or dialing again if the connection disrupts.

Resource recycling

We have already seen how to implement resource recycling, with a buffered channel with a pool of workers, in the previous chapter. There will be two methods as follows:

- A `Get` method that tries to receive a message from the channel or return a new instance.
- A `Put` method that tries to return an instance back to a channel or discard it.

This is a simple implementation of a pool with channels:

```
type A struct{}

type Pool chan *A

func (p Pool) Get() *A {
    select {
    case a := <-p:
        return a
    default:
        return new(A)
    }
}

func (p Pool) Put(a *A) {
    select {
    case p <- a:
    default:
    }
}
```

We can improve this using the `sync.Pool` structure, which implements a thread-safe set of objects that can be saved or retrieved. The only thing that needs to be defined is the behavior of the pool when creating a new object:

```
type Pool struct {
    // New optionally specifies a function to generate
    // a value when Get would otherwise return nil.
```

```
        // It may not be changed concurrently with calls to Get.
        New func() interface{}
        // contains filtered or unexported fields
    }
```

The pool offers two methods: `Get` and `Put`. These methods return an object from the pool (or create a new one) and place the object back in the pool. Since the `Get` method returns an `interface{}`, the value needs to be cast to the specific type in order to be used correctly. We talked extensively about buffer recycling and in the following example, we will try to implement one using `sync.Pool`.

We will need to define the pool and functions to obtain and release new buffers. Our buffers will have an initial capacity of 4 KB, and the `Put` function will ensure that the buffer is reset before putting it back in the pool, as shown in the following code example:

```
var pool = sync.Pool{
    New: func() interface{} {
        return bytes.NewBuffer(make([]byte, 0, 4096))
    },
}

func Get() *bytes.Buffer {
    return pool.Get().(*bytes.Buffer)
}

func Put(b *bytes.Buffer) {
    b.Reset()
    pool.Put(b)
}
```

Now we will create a series of goroutines, which will use the `WaitGroup` to signal when they're done, and will do the following:

- Wait a certain amount of time (1-5 seconds).
- Acquire a buffer.
- Write information on the buffer.
- Copy the content to the standard output.
- Release the buffer.

We will use a sleep time equal to 1 second, plus another second every 4 iterations of the loop, up to 5:

```
start := time.Now()
wg := sync.WaitGroup{}
wg.Add(20)
for i := 0; i < 20; i++ {
    go func(v int) {
        time.Sleep(time.Second * time.Duration(1+v/4))
        b := Get()
        defer func() {
            Put(b)
            wg.Done()
        }()
        fmt.Fprintf(b, "Goroutine %2d using %p, after %.0fs\n", v, b,
time.Since(start).Seconds())
        fmt.Printf("%s", b.Bytes())
    }(i)
}
wg.Wait()
```

The information in print also contains the buffer memory address. This will help us to confirm that the buffers are always the same and no new ones are created.

Slices recycling issues

With data structure with an underlying byte slice, such as `bytes.Buffer`, we should be careful when using them combined with `sync.Pool` or a similar mechanism of recycling. Let's change the previous example and collect the buffer's bytes instead of printing them to standard output. The following is an example code for this:

```
var (
    list = make([][]byte, 20)
    m sync.Mutex
)
for i := 0; i < 20; i++ {
    go func(v int) {
        time.Sleep(time.Second * time.Duration(1+v/4))
        b := Get()
        defer func() {
            Put(b)
            wg.Done()
        }()
        fmt.Fprintf(b, "Goroutine %2d using %p, after %.0fs\n", v, b,
time.Since(start).Seconds())
        m.Lock()
```

```
        list[v] = b.Bytes()
        m.Unlock()
    }(i)
}
wg.Wait()
```

So, what happens when we print the list of byte slices? We can see this in the following example:

```
for i := range list {
    fmt.Printf("%d - %s", i, list[i])
}
```

We get an unexpected result as the buffers have been overwritten. That's because the buffers are reusing the same underlying slice and overriding the content with every new usage.

A solution to this problem is usually to execute a copy of the bytes, instead of just assigning them:

```
m.Lock()
list[v] = make([]byte, b.Len())
copy(list[v], b.Bytes())
m.Unlock()
```

Conditions

In concurrent programming, a condition variable is a synchronization mechanism that contains threads waiting for the same condition to verify. In Go, this means that there are some goroutines waiting for something to occur. We already did an implementation of this using channels with a single goroutine waiting, as shown in the following example:

```
ch := make(chan struct{})
go func() {
    // do something
    ch <- struct{}{}
}()
go func() {
    // wait for condition
    <-ch
    // do something else
}
```

This approach is limited to a single goroutine, but it can be improved to support more listeners switching from message-sending to closing down the channel:

```
go func() {
    // do something
    close(ch)
}()
for i := 0; i < n; i++ {
    go func() {
        // wait for condition
        <-ch
        // do something else
    }()
}
```

Closing the channel works for more than one listener, but it does not allow them to use the channel any further after it closes.

The `sync.Cond` type is a tool that makes it possible to handle all this behavior in a better way. It uses a locker in its implementation and exposes three methods:

- `Broadcast`: This wakes all goroutines waiting for the condition.
- `Signal`: This wakes a single goroutine waiting for the condition, if there is at least one.
- `Wait`: This unlocks the locker, suspends execution of the goroutine, and later resumes the execution and locks it again, waiting for a `Broadcast` or `Signal`.

It is not required, but the `Broadcast` and `Signal` operations can be done while holding the locker, locking it before and releasing it after. The `Wait` method requires holding the locker before calling and unlocking it after the condition has been used.

Let's create a concurrent application which uses `sync.Cond` to orchestrate more goroutines. We will have a prompt from the command line, and each record will be written to a series of files. We will have a main structure that holds all the data:

```
type record struct {
    sync.Mutex
    buf string
    cond *sync.Cond
    writers []io.Writer
}
```

The condition we will be monitoring is a change in the buf field. In the Run method, the record structure will start several goroutines, one for each writer. Each goroutine will be waiting for the condition to trigger and will write in its file:

```go
func (r *record) Run() {
    for i := range r.writers {
        go func(i int) {
            for {
                r.Lock()
                r.cond.Wait()
                fmt.Fprintf(r.writers[i], "%s\n", r.buf)
                r.Unlock()
            }
        }(i)
    }
}
```

We can see that we lock the condition before using Wait, and we unlock it after using the value that our condition refers to. The main function will create a record and a series of files, according to the command-line arguments provided:

```go
// let's make sure we have at least a file argument
if len(os.Args) < 2 {
    log.Fatal("Please specify at least a file")
}
r := record{
    writers: make([]io.Writer, len(os.Args)-1),
}
r.cond = sync.NewCond(&r)
for i, v := range os.Args[1:] {
    f, err := os.Create(v)
    if err != nil {
        log.Fatal(err)
    }
    defer f.Close()
    r.writers[i] = f
}
r.Run()
```

We will then use bufio.Scanner to read lines and broadcast the change of the buf field. We will also accept a special value, \q, as a quit command:

```go
scanner := bufio.NewScanner(os.Stdin)
for {
    fmt.Printf(":> ")
    if !scanner.Scan() {
        break
```

```
    }
    r.Lock()
    r.buf = scanner.Text()
    r.Unlock()
    switch {
    case r.buf == `\q`:
        return
    default:
        r.cond.Broadcast()
    }
}
```

We can see that the change of `buf` is done while holding the lock and this is followed by the call to `Broadcast`, which wakes up all the goroutines waiting for the condition.

Synchronized maps

Built-in maps in Go are not thread-safe, and, therefore, trying to write from different goroutines can cause a runtime error: `concurrent map writes`. We can verify this using a simple program that tries to make changes concurrently:

```
func main() {
    var m = map[int]int{}
    wg := sync.WaitGroup{}
    wg.Add(10)
    for i := 0; i < 10; i++ {
        go func(i int) {
            m[i%5]++
            fmt.Println(m)
            wg.Done()
        }(i)
    }
    wg.Wait()
}
```

Reading while writing is also a runtime error, `concurrent map iteration and map write`, which we can see by running the following example:

```
func main() {
    var m = map[int]int{}
    var done = make(chan struct{})
    go func() {
        for i := 0; i < 100; i++ {
            time.Sleep(time.Nanosecond)
            m[i]++
        }
```

```
            close(done)
    }()
    for {
        time.Sleep(time.Nanosecond)
        fmt.Println(len(m), m)
        select {
        case <-done:
            return
        default:
        }
    }
}
```

Sometimes, trying to iterate a map (as the `Print` statement does) can cause panic such as `index out of range`, because the internal slices may have been allocated somewhere else.

A very easy strategy to make a map concurrent is to couple it with `sync.Mutex` or `sync.RWMutex`. This makes it possible to lock the map when executing the operations:

```
type m struct {
    sync.Mutex
    m map[int]int
}
```

We use the map for getting or setting the value, such as the following, for instance:

```
func (m *m) Get(key int) int {
    m.Lock()
    a := m.m[key]
    m.Unlock()
    return a
}

func (m *m) Put(key, value int) {
    m.Lock()
    m.m[key] = value
    m.Unlock()
}
```

We can also pass a function that takes a key-value pair and executes it for each tuple, while locking the map:

```
func (m *m) Range(f func(k, v int)) {
    m.Lock()
    for k, v := range m.m {
        f(k, v)
    }
    m.Unlock()
}
```

Go 1.9 introduced a structure called `sync.Map` that does exactly this. It is a very generic `map[interface{}]interface{}`, which makes it possible to execute thread-safe operations using the following methods:

- `Load`: Gets a value from the map for the given key.
- `Store`: Sets a value in the map for the given key.
- `Delete`: Removes the entry for the given key from the map.
- `LoadOrStore`: Returns the value for the key, if present, or the stored value.
- `Range`: Calls a function that returns a Boolean for each key-value pair in the map. The iteration stops if `false` is returned.

We can see how this works in the following snippet, in which we try to attempt several writes at the same time:

```
func main() {
    var m = sync.Map{}
    var wg = sync.WaitGroup{}
    wg.Add(1000)
    for i := 0; i < 1000; i++ {
        go func(i int) {
            m.LoadOrStore(i, i)
            wg.Done()
        }(i)
    }
    wg.Wait()
    i := 0
    m.Range(func(k, v interface{}) bool {
        i++
        return true
    })
    fmt.Println(i)
}
```

This application, unlike the version with a regular Map, does not crash and executes all the operations.

Semaphores

In the previous chapter, we saw how it is possible to use channels to create weighted semaphores. There is a better implementation in the experimental sync package. This can be found at: golang.org/x/sync/semaphore.

This implementation makes it possible to create a new semaphore, specifying the weight with semaphore.NewWeighted.

Quotas can be acquired using the Acquire method, specifying how many quotas you want to acquire. These can be released using the Release method, as shown in the following example:

```go
func main() {
    s := semaphore.NewWeighted(int64(10))
    ctx := context.Background()
    for i := 0; i < 20; i++ {
        if err := s.Acquire(ctx, 1); err != nil {
            log.Fatal(err)
        }
        go func(i int) {
            fmt.Println(i)
            s.Release(1)
        }(i)
    }
    time.Sleep(time.Second)
}
```

Acquiring quotas requires another argument besides the number, which is context.Context. This is another concurrency tool available in Go and we are going to see how to use this in the next chapter.

Atomic operations

The sync package delivers synchronization primitives, and, under the hood, it is using thread-safe operations on integers and pointers. We can find these functionalities in another package called sync/atomic, which can be used to create tools specific to the user use case, with better performance and less memory usage.

Integer operations

There is a series of functions for pointers to the different types of integers:

- `int32`
- `int64`
- `uint32`
- `uint64`
- `uintptr`

This includes a specific type of integer that represents a pointer, `uintptr`. The operation available for these types are as follows:

- `Load`: Retrieves the integer value from the pointer
- `Store`: Stores the integer value in the pointer
- `Add`: Adds the specified delta to the pointer value
- `Swap`: Stores a new value in the pointer and returns the old one
- `CompareAndSwap`: Swaps the new value for the old one only if this is the same as the specified one

clicker

This function can be very helpful for defining thread-safe components really easily. A very obvious example could be a simple integer counter that uses `Add` to change the counter, `Load` to retrieve the current value, and `Store` to reset it:

```
type clicker int32

func (c *clicker) Click() int32 {
    return atomic.AddInt32((*int32)(c), 1)
}

func (c *clicker) Reset() {
    atomic.StoreInt32((*int32)(c), 0)
}

func (c *clicker) Value() int32 {
    return atomic.LoadInt32((*int32)(c))
}
```

We can see it in action in a simple program, which tries to read, write, and reset the counter concurrently.

We define the `clicker` and `WaitGroup` and add the correct number of elements to the wait group as follows:

```
c := clicker(0)
wg := sync.WaitGroup{}
// 2*iteration + reset at 5
wg.Add(21)
```

We can launch a bunch of goroutines doing different actions, such as: 10 reads, 10 adds, and a reset:

```
for i := 0; i < 10; i++ {
    go func() {
        c.Click()
        fmt.Println("click")
        wg.Done()
    }()
    go func() {
        fmt.Println("load", c.Value())
        wg.Done()
    }()
    if i == 0 || i%5 != 0 {
        continue
    }
    go func() {
        c.Reset()
        fmt.Println("reset")
        wg.Done()
    }()
}
wg.Wait()
```

We will see the clicker acting as it is supposed to, executing concurrent sums without race conditions.

Thread-safe floats

The `atomic` package offers only primitives for integers, but since `float32` and `float64` are stored in the same data structure that `int32` and `int64` use, we use them to create an atomic float value.

The trick is to use the `math.Floatbits` functions to get the representation of a float as an unsigned integer and the `math.Floatfrombits` functions to transform an unsigned integer to a float. Let's see how this works with a `float64`:

```
type f64 uint64

func uf(u uint64) (f float64) { return math.Float64frombits(u) }
func fu(f float64) (u uint64) { return math.Float64bits(f) }

func newF64(f float64) *f64 {
    v := f64(fu(f))
    return &v
}

func (f *f64) Load() float64 {
   return uf(atomic.LoadUint64((*uint64)(f)))
}

func (f *f64) Store(s float64) {
   atomic.StoreUint64((*uint64)(f), fu(s))
}
```

Creating the `Add` function is a little bit more complicated. We need to get the value with `Load`, then compare and swap. Since this operation could fail because the load is an `atomic` operation and **compare and swap (CAS)** is another, we keep trying it until it succeeds in a loop:

```
func (f *f64) Add(s float64) float64 {
    for {
        old := f.Load()
        new := old + s
        if f.CompareAndSwap(old, new) {
            return new
        }
    }
}

func (f *f64) CompareAndSwap(old, new float64) bool {
    return atomic.CompareAndSwapUint64((*uint64)(f), fu(old), fu(new))
}
```

Thread-safe Boolean

We can also use `int32` to represent a Boolean value. We can use the integer `0` as `false`, and `1` as `true`, creating a thread-safe Boolean condition:

```
type cond int32

func (c *cond) Set(v bool) {
    a := int32(0)
    if v {
        a++
    }
    atomic.StoreInt32((*int32)(c), a)
}

func (c *cond) Value() bool {
    return atomic.LoadInt32((*int32)(c)) != 0
}
```

This will allow us to use the `cond` type as a thread-safe Boolean value.

Pointer operations

Pointer variables in Go are stored in `intptr` variables, integers large enough to hold a memory address. The `atomic` package makes it possible to execute the same operations for other integers types. There is a package that allows unsafe pointer operations, which offers the `unsafe.Pointer` type that is used in atomic operations.

In the following example, we define two integer variables and their relative integer pointers. Then we execute a swap of the first pointer with the second:

```
v1, v2 := 10, 100
p1, p2 := &v1, &v2
log.Printf("P1: %v, P2: %v", *p1, *p2)
atomic.SwapPointer((*unsafe.Pointer)(unsafe.Pointer(&p1)),
unsafe.Pointer(p2))
log.Printf("P1: %v, P2: %v", *p1, *p2)
v1 = -10
log.Printf("P1: %v, P2: %v", *p1, *p2)
v2 = 3
log.Printf("P1: %v, P2: %v", *p1, *p2)
```

After the swap, both pointers are now referring to the second variable; any change to the first value does not influence the pointers. Changing the second variable changes the value referred to by the pointers.

Value

The simplest tool we can use is `atomic.Value`. This holds `interface{}` and makes it possible to read and write it with thread safety. It exposes two methods, `Store` and `Load`, which make it possible to set or retrieve the value. As it happens, for other thread-safe tools, `sync.Value` must not be copied after its first use.

We can try to have many goroutines to set and read the same value. Each load operation gets the latest stored value and there are no errors being raised by concurrency:

```
func main() {
    var (
        v atomic.Value
        wg sync.WaitGroup
    )
    wg.Add(20)
    for i := 0; i < 10; i++ {
        go func(i int) {
            fmt.Println("load", v.Load())
            wg.Done()
        }(i)
        go func(i int) {
            v.Store(i)
            fmt.Println("store", i)
            wg.Done()
        }(i)
    }
    wg.Wait()
}
```

This is a very generic container; it can be used for any type of variable and the variable type should change from one to another. If the concrete type changes, it will make the method panic; the same thing applies to a `nil` empty interface.

Under the hood

The `sync.Value` type stores its data in a non-exported interface, as shown by the source code:

```
type Value struct {
    v interface{}
}
```

It uses a type of `unsafe` package to convert that structure into another one, which has the same data structure as an interface:

```
type ifaceWords struct {
    typ unsafe.Pointer
    data unsafe.Pointer
}
```

Two types with the same exact memory layout can be converted in this way, skipping the Go's type safety. This makes it possible to use `atomic` operations with the pointers and execute thread-safe `Store` and `Load` operations.

To get the lock for writing values, `atomic.Value` uses a compare and swap operation with the `unsafe.Pointer(^uintptr(0))` value (which is `0xffffffff`) in the type; it changes the value and replaces the type with the correct one.

In the same way, the load operation loops until the type is different to `0xffffffff`, before trying to read the value.

Using this expedient, `atomic.Value` is capable of storing and loading any value using other `atomic` operations.

Summary

In this chapter, we saw the tools that are available in the Go standard package for synchronization. They are located in two packages: `sync`, which provides high-level tools such as mutexes, and `sync/atomic`, which executes low-level operations.

First, we saw how to synchronize data using lockers. We saw how to use `sync.Mutex` to lock a resource regardless of the operation type, and `sync.RWMutex` to allow for concurrent readings and blocking writes. We should be careful using the second one because writes could be delayed by consecutive readings.

Next, we saw how to keep track of running operations in order to wait for the end of a series of goroutines, using `sync.WaitGroup`. This acts as a thread-safe counter for current goroutines and makes it possible to put the current goroutine to sleep until it reaches zero, using the `Wait` method.

Furthermore, we checked the `sync.Once` structure used to execute a functionality once, which allows the implementation of a thread-safe singleton, for instance. Then we used `sync.Pool` to reuse instances instead of creating new ones when possible. The only thing that a pool needs is the function that returns the new instance.

The `sync.Condition` struct represents a specific condition and uses a locker to change it, allowing a goroutine to wait for the change. This can be delivered to a single goroutine using `Signal`, or to all goroutines using `Broadcast`. The package also offers a thread-safe version of `sync.Map`.

Finally, we checked out the functionalities of `atomic`, which are mostly integer thread-safe operations: loading, saving, adding, swapping, and CAS. We saw also `atomic.Value`, which that makes it possible to change the value of an interface concurrently and does not allow it to change type after the first change.

The next chapter will be about the latest element introduced in Go concurrency: `Context`, which is an interface that handles deadlines, cancellations, and much more.

Questions

1. What's a race condition?
2. What happens when you try to execute read and write operations concurrently with a map?
3. What's the difference between `Mutex` and `RWMutex`?
4. Why are wait groups useful?
5. What's the main use of `Once`?
6. How can you use a `Pool`?
7. What's the advantage of using atomic operations?

13
Coordination Using Context

This chapter is about the relatively new context package and its usage in concurrent programming. It is a very powerful tool by defining a unique interface that's used in many different places in the standard library, as well as in many third-party packages.

The following topics will be covered in this chapter:

- Understanding what context is
- Studying its usage in the standard library
- Creating a package that uses context

Technical requirements

This chapter requires Go to be installed and your favorite editor to be set up. For more information, refer to Chapter 3, *An Overview of Go*.

Understanding context

Context is a relatively new component that entered the standard library in version 1.7. It is an interface for synchronization between goroutines that was used internally by the Go team and ended up being a core part of the language.

The interface

The main entity in the package is `Context` itself, which is an interface. It has only four methods:

```
type Context interface {
    Deadline() (deadline time.Time, ok bool)
    Done() <-chan struct{}
    Err() error
    Value(key interface{}) interface{}
}
```

Let's learn about these four methods here:

- `Deadline`: Returns the time when the context should be cancelled, together with a Boolean that is `false` when there is no deadline

- `Done`: Returns a receive-only channel of empty structs, which signals when the context should be cancelled

- `Err`: Returns `nil` while the `done` channel is open; otherwise it returns the cause of the context cancellation

- `Value`: Returns a value associated with a key for the current context, or `nil` if there's no value for the key

Context has many methods compared to the other interfaces of the standard library, which usually have one or two methods. Three of them are closely related:

- `Deadline` is the time for cancelling
- `Done` signals when the context is done
- `Err` returns the cause of the cancellation

The last method, `Value`, returns the value associated with a certain key. The rest of the package is a series of functions that allow you to create different types of contexts. Let's go through the various functions that comprise the package and look at various tools for creating and decorating contexts.

Default contexts

The TODO and Background functions return context.Context without the need for any input argument. The value that's returned is an empty context, though, their distinction is just semantic.

Background

Background is an empty context that doesn't get cancelled, hasn't got a deadline, and doesn't hold any values. It is mostly used by the main function as the root context or for testing purposes. The following is some example code for this context:

```
func main() {
    ctx := context.Background()
    done := ctx.Done()
    for i :=0; ;i++{
        select {
        case <-done:
            return
        case <-time.After(time.Second):
            fmt.Println("tick", i)
        }
    }
}
```

The full example is available here: https://play.golang.org/p/y_3ip7sdPnx.

We can see that, in the context of the example, the loop goes on infinitely because the context is never completed.

TODO

TODO is another empty context that should be used when the scope of the context isn't clear or if the type of context isn't available yet. It is used in the exact same way as Background. As a matter of fact, under the hood, they are the same thing; the distinction is only semantical. If we look at the source code, they have the exact same definition:

```
var (
    background = new(emptyCtx)
    todo = new(emptyCtx)
)
```

The source for this code can be found at `https://golang.org/pkg/context/?m=all#pkg-variables`.

These basic contexts can be extended using the other functions of the package. They will act as decorators and add more capabilities to them.

Cancellation, timeout, and deadline

The context we looked at is never cancelled, but the package offers different options for adding this functionality.

Cancellation

The `context.WithCancel` decorator function gets a context and returns another context and a function called `cancel`. The returned context will be a copy of the context that has a different `done` channel (the channel that marks that the current context is done) that gets closed when the parent context does or when the `cancel` function is called – whatever happens first.

In the following example, we can see that we wait a few seconds before calling the `cancel` function, and the program terminates correctly. The value of `Err` is the `context.Canceled` variable:

```
func main() {
    ctx, cancel := context.WithCancel(context.Background())
    time.AfterFunc(time.Second*5, cancel)
    done := ctx.Done()
    for i := 0; ; i++ {
        select {
        case <-done:
            fmt.Println("exit", ctx.Err())
            return
        case <-time.After(time.Second):
            fmt.Println("tick", i)
        }
    }
}
```

The full example is available here: `https://play.golang.org/p/fNHLIZL8eOL`.

Deadline

`context.WithDeadline` is another decorator, which specifies a time deadline as `time.Time`, and applies it to another context. If there is already a deadline and it is earlier than the one provided, the specified one gets ignored. If the `done` channel is still open when the deadline is met, it gets closed automatically.

In the following example, we set the deadline to be 5 seconds from now and call `cancel` 10 seconds after. The deadline arrives before the cancellation and `Err` returns a `context.DeadlineExceeded` error:

```
func main() {
    ctx, cancel := context.WithDeadline(context.Background(),
        time.Now().Add(5*time.Second))
    time.AfterFunc(time.Second*10, cancel)
    done := ctx.Done()
    for i := 0; ; i++ {
        select {
        case <-done:
            fmt.Println("exit", ctx.Err())
            return
        case <-time.After(time.Second):
            fmt.Println("tick", i)
        }
    }
}
```

The full example is available here: `https://play.golang.org/p/iyuOmd__CGH`.

We can see that the preceding example behaves exactly as expected. It will print the `tick` statement each second a few times until the the deadline is met and the error is returned.

Timeout

The last cancel-related decorator is `context.WithTimeout`, which allows you to specify a `time.Duration` together with the context and closes the `done` channel automatically when the timeout is passed.

If there a deadline active, the new value applies only if it's earlier than the parent. We can look at a pretty identical example, beside the context definition, and get the same result that we got for the deadline example:

```
func main() {
    ctx, cancel := context.WithTimeout(context.Background(),5*time.Second)
    time.AfterFunc(time.Second*10, cancel)
    done := ctx.Done()
    for i := 0; ; i++ {
        select {
        case <-done:
            fmt.Println("exit", ctx.Err())
            return
        case <-time.After(time.Second):
            fmt.Println("tick", i)
        }
    }
}
```

The full example is available here: `https://play.golang.org/p/-Zp63_e0zYD`.

Keys and values

The `context.WithValue` function creates a copy of the parent context that has the given key associated with the specified value. Its scope holds values that are relative to a single request while it gets processed and should not be used for other scopes, such as optional function parameters.

The key should be something that can be compared, and it's a good idea to avoid `string` values because two different packages using context could overwrite each other's values. The suggestion is to use user-defined concrete types such as `struct{}`.

Here, we can see an example where we take a base context and we add a different value for each goroutine, using an empty struct as a key:

```
type key struct{}

type key struct{}

func main() {
    ctx, canc := context.WithCancel(context.Background())
    wg := sync.WaitGroup{}
    wg.Add(5)
    for i := 0; i < 5; i++ {
        go func(ctx context.Context) {
```

```
            v := ctx.Value(key{})
            fmt.Println("key", v)
            wg.Done()
            <-ctx.Done()
            fmt.Println(ctx.Err(), v)
        }(context.WithValue(ctx, key{}, i))
    }
    wg.Wait()
    canc()
    time.Sleep(time.Second)
}
```

The full example is available here: `https://play.golang.org/p/lM61u_QKEW1`.

We can also see that cancelling the parent cancels the other contexts. Another valid key type could be exported pointer values, which won't be the same, even if the underlying data is:

```
type key *int

func main() {
    k := new(key)
    ctx, canc := context.WithCancel(context.Background())
    wg := sync.WaitGroup{}
    wg.Add(5)
    for i := 0; i < 5; i++ {
        go func(ctx context.Context) {
            v := ctx.Value(k)
            fmt.Println("key", v, ctx.Value(new(key)))
            wg.Done()
            <-ctx.Done()
            fmt.Println(ctx.Err(), v)
        }(context.WithValue(ctx, k, i))
    }
    wg.Wait()
    canc()
    time.Sleep(time.Second)
}
```

The full example is available here: `https://play.golang.org/p/05XJwWF0-0n`.

We can see that defining a key pointer with the same underlying value doesn't return the expected value.

Context in the standard library

Now that we've covered the contents of the package, we will look at how to use them with the standard package or in an application. Context is used in a few functions and methods of standard packages, mostly network packages. Let's go over them now:

- `http.Server` uses it with the `Shutdown` method so that it has full control over timeout or to cancel an operation.
- `http.Request` allows you to set a context using the `WithContext` method. It also allows you to get the current context using `Context`.
- In the `net` package, `Listen`, `Dial`, and `Lookup` have a version that uses `Context` to control deadlines and timeouts.
- In the `database/sql` package, context is used to stop or timeout many different operations.

HTTP requests

Before the introduction of the official package, each HTTP-related framework was using its own version of context to store data relative to HTTP requests. This resulted in fragmentation, and the reuse of handlers and middleware wasn't possible without rewriting the middleware or any specific binding code.

Passing scoped values

The introduction of `context.Context` in `http.Request` tries to address this issue by defining a single interface that can be assigned, recovered, and used in various handlers.

The downside is that a context isn't assigned automatically to a request, and context values cannot be recycled. There should be no really good reason to do that since the context should store data that's specific to a certain package or scope, and the packages themselves should be the only ones that are able to interact with them.

A good pattern is the usage of a unique unexported key type combined with auxiliary functions to get or set a certain value:

```
type keyType struct{}

var key = &keyType{}

func WithKey(ctx context.Context, value string) context.Context {
    return context.WithValue(ctx, key, value)
```

```
}

func GetKey(ctx context.Context) (string, bool) {
    v := ctx.Value(key)
    if v == nil {
        return "", false
    }
    return v.(string), true
}
```

A context request is the only case in the standard library where it is stored in the data structure with the `WithContext` method and it's accessed using the `Context` method. This has been done in order to not break the existing code, and maintain the promise of compatibility of Go 1.

The full example is available here: `https://play.golang.org/p/W6gGp_InoMp`.

Request cancellation

A good usage of context is for cancellation and timeout when you're executing an HTTP request using `http.Client`, which handles the interruption automatically from the context. The following example does exactly that:

```
func main() {
    const addr = "localhost:8080"
    http.HandleFunc("/", func(w http.ResponseWriter, r *http.Request) {
        time.Sleep(time.Second * 5)
    })
    go func() {
        if err := http.ListenAndServe(addr, nil); err != nil {
            log.Fatalln(err)
        }
    }()
    req, _ := http.NewRequest(http.MethodGet, "http://"+addr, nil)
    ctx, canc := context.WithTimeout(context.Background(), time.Second*2)
    defer canc()
    time.Sleep(time.Second)
    if _, err := http.DefaultClient.Do(req.WithContext(ctx)); err != nil {
        log.Fatalln(err)
    }
}
```

The context cancellation method can also be used to interrupt the current HTTP request that's passed to a client. In a scenario where we are calling different endpoints and returning the first result that's received, it would be a good idea to cancel the other requests.

Let's create an application that runs a query on different search engines and returns the results from the quickest one, cancelling the others. We can create a web server that has a unique endpoint that answers back in 0 to 10 seconds:

```go
const addr = "localhost:8080"
http.HandleFunc("/", func(w http.ResponseWriter, r *http.Request) {
    d := time.Second * time.Duration(rand.Intn(10))
    log.Println("wait", d)
    time.Sleep(d)
})
go func() {
    if err := http.ListenAndServe(addr, nil); err != nil {
        log.Fatalln(err)
    }
}()
```

We can use a cancellable context for the requests, combined with a wait group to synchronize it with the end of the request. Each goroutine will create a request and try to send the result using a channel. Since we are only interested in the first one, we will use `sync.Once` to limit it:

```go
ctx, canc := context.WithCancel(context.Background())
ch, o, wg := make(chan int), sync.Once{}, sync.WaitGroup{}
wg.Add(10)
for i := 0; i < 10; i++ {
    go func(i int) {
        defer wg.Done()
        req, _ := http.NewRequest(http.MethodGet, "http://"+addr, nil)
        if _, err := http.DefaultClient.Do(req.WithContext(ctx)); err != nil {
            log.Println(i, err)
            return
        }
        o.Do(func() { ch <- i })
    }(i)
}
log.Println("received", <-ch)
canc()
log.Println("cancelling")
wg.Wait()
```

When this program runs, we will see that one of the requests is completed successfully and gets sent to the channel, while the others are either cancelled or ignored.

HTTP server

The `net/http` package has several uses of context, including stopping the listener or being part of a request.

Shutdown

`http.Server` allows us to pass a context for the shutdown operation. This allows to us to use some of the context capabilities, such as cancelling and timeout. We can define a new server with its `mux` and a cancellable context:

```
mux := http.NewServeMux()
server := http.Server{
    Addr: ":3000",
    Handler: mux,
}
ctx, canc := context.WithCancel(context.Background())
defer canc()
mux.HandleFunc("/shutdown", func(w http.ResponseWriter, r *http.Request) {
    w.Write([]byte("OK"))
    canc()
})
```

We can launch the server in a separate goroutine:

```
go func() {
    if err := server.ListenAndServe(); err != nil {
        if err != http.ErrServerClosed {
            log.Fatal(err)
        }
    }
}()
```

The context will be complete when the shutdown endpoint is called and the cancellation function is invoked. We can wait for that event and then use another context with a timeout to call the shutdown method:

```
select {
case <-ctx.Done():
    ctx, canc := context.WithTimeout(context.Background(), time.Second*5)
    defer canc()
```

```
    if err := server.Shutdown(ctx); err != nil {
        log.Fatalln("Shutdown:", err)
    } else {
        log.Println("Shutdown:", "ok")
    }
}
```

This will allow us to terminate the server effectively within the timeout, after which it will terminate with an error.

Passing values

Another usage of context in a server is as a propagation of values and cancellation between different HTTP handlers. Let's look at an example where each request has a unique key that is an integer. We will use a couple of functions that are similar to the example where we had values using integers. The generation of a new key will be done with atomic:

```
type keyType struct{}

var key = &keyType{}

var counter int32

func WithKey(ctx context.Context) context.Context {
    return context.WithValue(ctx, key, atomic.AddInt32(&counter, 1))
}

func GetKey(ctx context.Context) (int32, bool) {
    v := ctx.Value(key)
    if v == nil {
        return 0, false
    }
    return v.(int32), true
}
```

Now, we can define another function that takes any HTTP handler and creates the context, if necessary, and adds the key to it:

```
func AssignKeyHandler(h http.Handler) http.HandlerFunc {
    return func(w http.ResponseWriter, r *http.Request) {
        ctx := r.Context()
        if ctx == nil {
            ctx = context.Background()
        }
        if _, ok := GetKey(ctx); !ok {
            ctx = WithKey(ctx)
```

```
        }
        h.ServeHTTP(w, r.WithContext(ctx))
    }
}
```

By doing this, we can define a very simple handler that serves files under a certain root directory. This function will use the key from the context to log information correctly. It will also check that the file exists before trying to serve it:

```
func ReadFileHandler(root string) http.HandlerFunc {
    root = filepath.Clean(root)
    return func(w http.ResponseWriter, r *http.Request) {
        k, _ := GetKey(r.Context())
        path := filepath.Join(root, r.URL.Path)
        log.Printf("[%d] requesting path %s", k, path)
        if !strings.HasPrefix(path, root) {
            http.Error(w, "not found", http.StatusNotFound)
            log.Printf("[%d] unauthorized %s", k, path)
            return
        }
        if stat, err := os.Stat(path); err != nil || stat.IsDir() {
            http.Error(w, "not found", http.StatusNotFound)
            log.Printf("[%d] not found %s", k, path)
            return
        }
        http.ServeFile(w, r, path)
        log.Printf("[%d] ok: %s", k, path)
    }
}
```

We can combine those handlers to serve content from different folders, such as the home user or the temporary directory:

```
home, err := os.UserHomeDir()
if err != nil {
    log.Fatal(err)
}
tmp := os.TempDir()
mux := http.NewServeMux()
server := http.Server{
    Addr: ":3000",
    Handler: mux,
}

mux.Handle("/tmp/", http.StripPrefix("/tmp/",
AssignKeyHandler(ReadFileHandler(tmp))))
mux.Handle("/home/", http.StripPrefix("/home/",
AssignKeyHandler(ReadFileHandler(home))))
```

```
if err := server.ListenAndServe(); err != nil {
    if err != http.ErrServerClosed {
        log.Fatal(err)
    }
}
```

We are using `http.StipPrefix` to remove the first part of the path and obtain the relative path, and pass it to the handler underneath. The resulting server will use context to pass the key value between handlers – this allows us to create another similar handler and use the `AssignKeyHandler` function to wrap the handler and `GetKey(r.Context())` to access the key inside our handler.

TCP dialing

The network package offers context-related functionalities, such as dialing cancellation when we're dialing or listening to incoming connections. It allows us to use the timeout and cancellation capabilities of context when dialing a connection.

Cancelling a connection

In order to test out the usage of context in a TCP connection, we can create a goroutine with a TCP server that will wait a period of time before starting the listener:

```
addr := os.Args[1]
go func() {
    time.Sleep(time.Second)
    listener, err := net.Listen("tcp", addr)
    if err != nil {
        log.Fatalln("Listener:", addr, err)
    }
    c, err := listener.Accept()
    if err != nil {
        log.Fatalln("Listener:", addr, err)
    }
    defer c.Close()
}()
```

We can use a context with a timeout that's lower than the server waiting time. We have to use `net.Dialer` in order to use the context in a dial operation:

```
ctx, canc := context.WithTimeout(context.Background(),
    time.Millisecond*100)
defer canc()
conn, err := (&net.Dialer{}).DialContext(ctx, "tcp", os.Args[1])
```

```
if err != nil {
    log.Fatalln("-> Connection:", err)
}
log.Println("-> Connection to", os.Args[1])
conn.Close()
```

The application will try to connect for a short time, but will eventually give up when the context expires, returning an error.

In a situation where you want to establish a single connection from a series of endpoints, context cancellation would be a perfect use case. All the connection attempts would share the same context, and the first connection that dials correctly would call the cancellation, stopping the other attempts. We will create a single server that is listening to one of the addresses we will try to call:

```
list := []string{
    "localhost:9090",
    "localhost:9091",
    "localhost:9092",
}
go func() {
    listener, err := net.Listen("tcp", list[0])
    if err != nil {
        log.Fatalln("Listener:", list[0], err)
    }
    time.Sleep(time.Second * 5)
    c, err := listener.Accept()
    if err != nil {
        log.Fatalln("Listener:", list[0], err)
    }
    defer c.Close()
}()
```

Then, we can try to dial all three addresses and cancel the context as soon as one connects. We will use a `WaitGroup` to synchronize with the end of the goroutines:

```
ctx, canc := context.WithTimeout(context.Background(), time.Second*10)
defer canc()
wg := sync.WaitGroup{}
wg.Add(len(list))
for _, addr := range list {
    go func(addr string) {
        defer wg.Done()
        conn, err := (&net.Dialer{}).DialContext(ctx, "tcp", addr)
        if err != nil {
            log.Println("-> Connection:", err)
            return
```

```
        }
        log.Println("-> Connection to", addr, "cancelling context")
        canc()
        conn.Close()
    }(addr)
}
wg.Wait()
```

What we will see in the output of this program is one connection succeeding, followed by the cancellation error of the other attempt.

Database operations

We aren't looking at the `sql/database` package in this book, but for the sake of completion, it is worth mentioning that it uses context too. Most of its operations have a counterpart with context, for instance:

- Beginning a new transaction
- Executing a query
- Pinging the database
- Preparing a query

This concludes the packages in the standard library that use context. Next, we are going to try to use context to build a package to allow the user of that package to cancel requests.

Experimental packages

A notable example in the experimental package that uses context is one we've already looked at – semaphore. Now that we have a better understanding of what context is for, it should be pretty obvious why the acquire operation also takes a context in as an argument.

When creating our application, we can provide a context with a timeout or cancellation and act accordingly:

```
func main() {
    s := semaphore.NewWeighted(int64(5))
    ctx, canc := context.WithTimeout(context.Background(), time.Second)
    defer canc()
    wg := sync.WaitGroup{}
    wg.Add(20)
    for i := 0; i < 20; i++ {
        go func(i int) {
```

```
            defer wg.Done()
            if err := s.Acquire(ctx, 1); err != nil {
                fmt.Println(i, err)
                return
            }
            go func(i int) {
                fmt.Println(i)
                time.Sleep(time.Second / 2)
                s.Release(1)
            }(i)
        }(i)
    }
    wg.Wait()
}
```

Running this application will show that the semaphore is acquired for the first second, but after that the context expires and all the remaining operations fail.

Context in your application

`context.Context` is the perfect tool to integrate into your package or application if it has operations that could take a long time and a user can cancel them, or if they should have time limitations such timeouts or deadlines.

Things to avoid

Even though the context scope has been made very clear by the Go team, developers have been using it in various ways – some less orthodox than others. Let's check out some of them and which alternatives there are, instead of resorting to context.

Wrong types as keys

The first practice to avoid is the usage of built-in types as keys. This is problematic because they can be overwritten because two interfaces with the same built-in values are considered the same, as shown in the following example:

```
func main() {
    var a interface{} = "request-id"
    var b interface{} = "request-id"
    fmt.Println(a == b)

    ctx := context.Background()
```

```
    ctx = context.WithValue(ctx, a, "a")
    ctx = context.WithValue(ctx, b, "b")
    fmt.Println(ctx.Value(a), ctx.Value(b))
}
```

The full example is available here: https://play.golang.org/p/2W3noYQP5eh.

The first print instruction outputs true, and since the keys are compared by value, the second assignment shadows the first, resulting in the values for both keys being the same. A potential solution to this is to use an empty struct custom type, or an unexported pointer to a built-in value.

Passing parameters

It might so happen that you need to travel a long way through a stack of function calls. A very tempting solution would be to use a context to store that value and recall it only in the function that needs it. It is generally not a good idea to hide a required parameter that should be passed explicitly. It results in less readable code because it won't make it clear what influences the execution of a certain function.

It is still much better to pass the function down the stack. If the parameters list is getting too long, then it could be grouped into one or more structs in order to be more readable.

Let's have a look at the following function:

```
func SomeFunc(ctx context.Context,
    name, surname string, age int,
    resourceID string, resourceName string) {}
```

The parameters could be grouped in the following way:

```
type User struct {
    Name string
    Surname string
    Age int
}

type Resource struct {
    ID string
    Name string
}

func SomeFunc(ctx context.Context, u User, r Resource) {}
```

Optional arguments

Context should be used to pass optional parameters around, and also used as a sort of catch-all, like Python `kwargs` or JavaScript `arguments`. Using context as a substitute for behaviors can be very problematic because it could cause the shadowing of variables, like we saw in the example of `context.WithValue`.

Another big drawback of this approach is hiding what's happening and making the code more obscure. A much better approach when it comes to optional values is using a pointer to structure arguments – this allows you to avoid passing the structure at all with `nil`.

Let's say you had the following code:

```
// This function has two mandatory args and 4 optional ones
func SomeFunc(ctx context.Context, arg1, arg2 int,
    opt1, opt2, opt3, opt4 string) {}
```

By using `Optional`, you would have something like this:

```
type Optional struct {
    Opt1 string
    Opt2 string
    Opt3 string
    Opt4 string
}

// This function has two mandatory args and 4 optional ones
func SomeFunc(ctx context.Context, arg1, arg2 int, o *Optional) {}
```

Globals

Some global variables can be stored in a context so that they can be passed through a series of function calls. This is generally not good practice since globals are available in every point of the application, so using context to store and recall them is pointless and a waste of resources and performance. If your package has some globals, you can use the Singleton pattern we looked at in `Chapter 12`, *Synchronization with sync and atomic,* to allow access to them from any point of your package or application.

Building a service with Context

We will now focus on how to create packages that support the usage of context. This will help us put together what we've learned about concurrency up until now. We will try to create a concurrent file search that makes uses of channels, goroutines, synchronization, and context.

Main interface and usage

The signature of the package will include a context, the root folder, the search term, and a couple of optional parameters:

- **Search in contents**: Will look for the string in the file's contents instead of the name
- **Exclude list**: Will not search the files with the selected name/names

The function would look something like this:

```
type Options struct {
    Contents bool
    Exclude []string
}

func FileSearch(ctx context.Context, root, term string, o *Options)
```

Since it should be a concurrent function, the return type could be a channel of result, which could be either an error or a series of matches in a file. Since we can search for the names of content, the latter could have more than one match:

```
type Result struct {
    Err error
    File string
    Matches []Match
}

type Match struct {
    Line int
    Text string
}
```

The previous function will return a receive-only channel of the Result type:

```
func FileSearch(ctx context.Context, root, term string, o *Options) <-chan
Result
```

Here, this function would keep receiving values from the channel until it gets closed:

```
for r := range FileSearch(ctx, directory, searchTerm, options) {
    if r.Err != nil {
        fmt.Printf("%s - error: %s\n", r.File, r.Err)
        continue
    }
    if !options.Contents {
        fmt.Printf("%s - match\n", r.File)
        continue
    }
    fmt.Printf("%s - matches:\n", r.File)
    for _, m := range r.Matches {
        fmt.Printf("\t%d:%s\n", m.Line, m.Text)
    }
}
```

Exit and entry points

The result channel should be closed by either the cancellation of the context, or by the search being over. Since a channel cannot be closed twice, we can use `sync.Once` to avoid closing the channel for the second time. To keep track of the goroutines that are running, we can use `sync.Waitgroup`:

```
ch, wg, once := make(chan Result), sync.WaitGroup{}, sync.Once{}
go func() {
    wg.Wait()
    fmt.Println("* Search done *")
    once.Do(func() {
        close(ch)
    })
}()
go func() {
    <-ctx.Done()
    fmt.Println("* Context done *")
    once.Do(func() {
        close(ch)
    })
}()
```

We could launch a goroutine for each file so that we can define a private function that we can use as an entry point and then use it recursively for subdirectories:

```
func fileSearch(ctx context.Context, ch chan<- Result, wg *sync.WaitGroup,
file, term string, o *Options)
```

The main exported function will start by adding a value to the wait group. It will then launch the private function, starting it as an asynchronous process:

```
wg.Add(1)
go fileSearch(ctx, ch, &wg, root, term, o)
```

The last thing each `fileSearch` should do is call `WaitGroup.Done` to mark the end of the current file.

Exclude list

The private function will decrease the wait group counter before it finishes using the `Done` method.. Besides that, the first thing it should do is check the filename so that it can skip it if it is in the exclusion list:

```
defer wg.Done()
_, name := filepath.Split(file)
if o != nil {
    for _, e := range o.Exclude {
        if e == name {
            return
        }
    }
}
```

If that is not the case, we can check the current file's information using `os.Stat` and send an error to the channel if we don't succeed. Since we cannot risk causing a panic by sending into a closed channel, we can check whether the context is done, and if not, send the error:

```
info, err := os.Stat(file)
if err != nil {
    select {
    case <-ctx.Done():
        return
    default:
        ch <- Result{File: file, Err: err}
    }
    return
}
```

Handling directories

The information that's received will tell us whether the file is a directory or not. If it is a directory, we can get a list of files and handle the error, as we did earlier with `os.Stat`. Then, we can launch another series of searches, one for each file, if the context isn't already done. The following code sums up these operations:

```
if info.IsDir() {
    files, err := ioutil.ReadDir(file)
    if err != nil {
        select {
        case <-ctx.Done():
            return
        default:
            ch <- Result{File: file, Err: err}
        }
        return
    }
    select {
    case <-ctx.Done():
    default:
        wg.Add(len(files))
        for _, f := range files {
            go fileSearch(ctx, ch, wg, filepath.Join(file,
        f.Name()), term, o)
        }
    }
    return
}
```

Checking file names and contents

If the file is a regular file and not a directory, we can compare the file name or its contents, depending on the options that are specified. Checking the file name is pretty easy:

```
if o == nil || !o.Contents {
    if name == term {
        select {
        case <-ctx.Done():
        default:
            ch <- Result{File: file}
        }
    }
    return
}
```

If we are searching for the contents, we should open the file:

```
f, err := os.Open(file)
if err != nil {
    select {
    case <-ctx.Done():
    default:
        ch <- Result{File: file, Err: err}
    }
    return
}
defer f.Close()
```

Then, we can read the file line by line to search for the selected term. If the context expires while we are reading the file, we will stop all operations:

```
scanner, matches, line := bufio.NewScanner(f), []Match{}, 1
for scanner.Scan() {
    select {
    case <-ctx.Done():
        break
    default:
        if text := scanner.Text(); strings.Contains(text, term) {
            matches = append(matches, Match{Line: line, Text: text})
        }
        line++
    }
}
```

Finally, we can check for errors from the scanner. If there's none and the search has results, we can send all the matches to the output channel:

```
select {
case <-ctx.Done():
    break
default:
    if err := scanner.Err(); err != nil {
        ch <- Result{File: file, Err: err}
        return
    }
    if len(matches) != 0 {
        ch <- Result{File: file, Matches: matches}
    }
}
```

In less than 200 lines, we created a concurrent file search function that uses one goroutine per file. It takes advantage of a channel to send results and synchronization primitives in order to coordinate operations.

Summary

In this chapter, we looked at what one of the newer packages, context, is all about. We saw that Context is a simple interface that has four methods, and should be used as the first argument of a function. Its main scope is to handle cancellation and deadlines to synchronize concurrent operations and provide the user with functionality to cancel an operation.

We saw how the default contexts, Background and TODO, don't allow cancellation, but they can be extended using various functions of the package to add timeouts or cancellation. We also talked about the capabilities of context when it comes to holding values and how this should be used carefully in order to avoid shadowing and other problems.

Then, we dived into the standard package to see where context is already used. This included the HTTP capabilities of requests, where it can be used for values, cancellation, and timeout, and the server shutdown operation. We also saw how the TCP package allows us to use it in a similar fashion with a practical example, and we also listed the operations in the database package that allow us to use context to cancel them.

Before building our own functionality using context, we went into some of the uses that should be avoided, from using the wrong types of keys to using context to pass values around that should be in a function or method signature instead. Then, we proceeded to create a function that searches files and contents, using what we have learned about concurrency from the last three chapters.

The next chapter will conclude the concurrency section of this book by showing off the most common Go concurrency patterns and their usage. This will allow us to put together all that we have learned so far about concurrency in some very common and effective configurations.

Questions

1. What is a context in Go?
2. What's the difference between cancellation, deadline, and timeout?
3. What are the best practices when passing values with a context?
4. Which standard packages already use context?

14
Implementing Concurrency Patterns

This chapter will be about concurrency patterns and how to use them to build robust system applications. We have already looked at all the tools that are involved in concurrency (goroutines and channels, `sync` and `atomic`, and context) so now we will look at some common ways of combining them in patterns so that we can use them in our programs.

The following topics will be covered in this chapter:

- Beginning with generators
- Sequencing with pipelines
- Muxing and demuxing
- Other patterns
- Resource leaking

Technical requirements

This chapter requires Go to be installed and your favorite editor to be set up. For more information, please refer to `Chapter 3`, *An Overview of Go*.

Beginning with generators

A generator is a function that returns the next value of a sequence each time it is called. The biggest advantage of using generators is the lazy creation of new values of the sequence. In Go, this can either be expressed with an interface or with a channel. One of the upsides of generators when they're used with channels is that they produce values concurrently, in a separate goroutine, leaving the main goroutine capable of executing other kinds of operations on them.

It can be abstracted with a very simple interface:

```
type Generator interface {
    Next() interface{}
}

type GenInt64 interface {
    Next() int64
}
```

The return type of the interface will depend on the use case, which is int64 in our case. The basic implementation of it could be something such as a simple counter:

```
type genInt64 int64

func (g *genInt64) Next() int64 {
    *g++
    return int64(*g)
}
```

This implementation is not thread-safe, so if we try to use it with goroutines, we could lose some elements on the way:

```
func main() {
    var g genInt64
    for i := 0; i < 1000; i++ {
        go func(i int) {
            fmt.Println(i, g.Next())
        }(i)
    }
    time.Sleep(time.Second)
}
```

A simple way to make a generator concurrent is to execute atomic operations on the integer.

This will make the concurrent generator thread-safe, with very few changes to the code needing to happen:

```
type genInt64 int64

func (g *genInt64) Next() int64 {
    return atomic.AddInt64((*int64)(g), 1)
}
```

This will avoid race conditions in the application. However, there is another implementation that's possible, but this requires the use of channels. The idea is to produce the value in a goroutine and then pass it in a shared channel to the next method, as shown in the following example:

```
type genInt64 struct {
    ch chan int64
}

func (g genInt64) Next() int64 {
    return <-g.ch
}

func NewGenInt64() genInt64 {
    g := genInt64{ch: make(chan int64)}
    go func() {
        for i := int64(0); ; i++ {
            g.ch <- i
        }
    }()
    return g
}
```

The loop will go on forever, and will be blocking in the send operation when the generator user stops requesting new values with the Next method.

The code was structured this way because we were trying to implement the interface we defined at the beginning. We could also just return a channel and use it for receiving:

```
func GenInt64() <-chan int64 {
  ch:= make(chan int64)
    go func() {
        for i := int64(0); ; i++ {
            ch <- i
        }
    }()
    return ch
}
```

The main advantage of using the channel directly is the possibility of including it in `select` statements in order to choose between different channel operations. The following shows a `select` between two different generators:

```go
func main() {
    ch1, ch2 := GenInt64(), GenInt64()
    for i := 0; i < 20; i++ {
        select {
        case v := <-ch1:
            fmt.Println("ch 1", v)
        case v := <-ch2:
            fmt.Println("ch 2", v)
        }
    }
}
```

Avoiding leaks

It is a good idea to allow the loop to end in order to avoid goroutine and resource leakage. Some of these issues are as follows:

- When a goroutine hangs without returning, the space in memory remains used, contributing to the application's size in memory. The goroutine and the variables it defines in the stack will get collected by the GC only when the goroutine returns or panics.
- If a file remains open, this can prevent other processes from executing operations on it. If the number of files that are open reaches the limit imposed by the OS, the process will not be able to open other files (or accept network connections).

An easy solution to this problem is to always use `context.Context` so that you have a well-defined exit point for the goroutine:

```go
func NewGenInt64(ctx context.Context) genInt64 {
    g := genInt64{ch: make(chan int64)}
    go func() {
        for i := int64(0); ; i++ {
            select {
            case g.ch <- i:
                // do nothing
            case <-ctx.Done():
                close(g.ch)
                return
            }
        }
```

```
    }()
    return g
}
```

This can be used to generate values until there is a need for them and cancel the context when there's no need for new values. The same pattern can be applied to a version that returns a channel. For instance, we could use the `cancel` function directly or set a timeout on the context:

```
func main() {
    ctx, cancel := context.WithTimeout(context.Background(),
time.Millisecond*100)
    defer cancel()
    g := NewGenInt64(ctx)
    for i := range g.ch {
        go func(i int64) {
            fmt.Println(i, g.Next())
        }(i)
    }
    time.Sleep(time.Second)
}
```

The generator will produce numbers until the context that's provided expires. At this point, the generator will close the channel.

Sequencing with pipelines

A pipeline is a way of structuring the application flow, and is obtained by splitting the main execution into stages that can talk with one another using certain means of communication. This could be either of the following:

- External, such as a network connection or a file
- Internal to the application, like Go's channels

The first stage is often referred to as the producer, while the last one is often called the consumer.

The set of concurrency tools that Go offers allows us to efficiently use multiple CPUs and optimize their usage by blocking input or output operations. Channels in particular are the perfect tools for internal pipeline communication. They can be represented by functions that receive an inbound channel and return an outbound one. The base structure would look something like this:

```
func stage(in <-chan interface{}) <-chan interface{} {
    var out = make(chan interface{})
    go func() {
        for v := range in {
            v = v.(int)+1 // some operation
            out <- v
        }
        close(out)
    }()
    return out
}
```

We create a channel of the same type as the input channel and return it. In a separate goroutine, we receive data from the input channel, perform an operation on the data, and we send it to the output channel.

This pattern can be further improved by the use of context.Context so that we have greater control of the application flow. It would look something like the following code:

```
func stage(ctx context.Context, in <-chan interface{}) <-chan interface{} {
    var out = make(chan interface{})
    go func() {
        defer close(out)
        for v := range in {
            v = v.(int)+1 // some operation
            select {
                case out <- v:
                case <-ctx.Done():
                    return
            }
        }
    }()
    return out
}
```

There are a couple of general rules that we should follow when designing pipelines:

- Intermediate stages will receive an inbound channel and return another one.
- The producer will not receive any channel, but return one.

- The consumer will receive a channel without returning one.
- Each stage will close the channel on creation when it's done sending messages.
- Each stage should keep receiving from the input channel until it's closed.

Let's create a simple pipeline that filters lines from a reader using a certain string and prints the filtered lines, highlighting the search string. We can start with the first stage – the source – which will not receive any channel in its signature but will use a reader to scan lines. We take a context in for reacting to an early exit request (context cancellation) and a bufio scanner to read line by line. The following code shows this:

```
func SourceLine(ctx context.Context, r io.ReadCloser) <-chan string {
    ch := make(chan string)
    go func() {
        defer func() { r.Close(); close(ch) }()
        s := bufio.NewScanner(r)
        for s.Scan() {
            select {
            case <-ctx.Done():
                return
            case ch <- s.Text():
            }
        }
    }()
    return ch
}
```

We can split the remaining operations into two phases: a filtering phase and a writing phase. The filtering phase will simply filter from a source channel to an output channel. We are still passing context in order to avoid sending extra data if the context is already complete. This is the text filter implementation:

```
func TextFilter(ctx context.Context, src <-chan string, filter string) <-chan string {
    ch := make(chan string)
    go func() {
        defer close(ch)
        for v := range src {
            if !strings.Contains(v, filter) {
                continue
            }
            select {
            case <-ctx.Done():
                return
            case ch <- v:
            }
        }
```

```
    } ()
    return ch
}
```

Finally, we have the final stage, the consumer, which will print the output in a writer and will also use a context for early exit:

```
func Printer(ctx context.Context, src <-chan string, color int, highlight
string, w io.Writer) {
    const close = "\x1b[39m"
    open := fmt.Sprintf("\x1b[%dm", color)
    for {
        select {
        case <-ctx.Done():
            return
        case v, ok := <-src:
            if !ok {
                return
            }
            i := strings.Index(v, highlight)
            if i == -1 {
                panic(v)
            }
            fmt.Fprint(w, v[:i], open, highlight, close,
v[i+len(highlight):], "\n")
        }
    }
}
```

The use of this function is as follows:

```
func main() {
    var search string
    ...
    ctx := context.Background()
    src := SourceLine(ctx, ioutil.NopCloser(strings.NewReader(sometext)))
    filter := TextFilter(ctx, src, search)
    Printer(ctx, filter, 31, search, os.Stdout)
}
```

With this approach, we learned how to split a complex operation into a simple task that's executed by stages and connected using channels.

Muxing and demuxing

Now that we are familiar with pipelines and stages, we can introduce two new concepts:

- **Muxing (multiplexing) or fan-out**: Receiving from one channel and sending to multiple channels
- **Demuxing (demultiplexing) or fan-in**: Receiving from multiple channels and sending through one channel

This pattern is very common and allows us to use the power of concurrency in different ways. The most obvious way is to distribute data from a channel that is quicker than its following step, and create more than one instance of such steps to make up for the difference in speed.

Fan-out

The implementation of multiplexing is pretty straightforward. The same channel needs to be passed to different stages so that each one will be reading from it.

Each goroutine is competing for resources during the runtime schedule, so if we want to reserve more of them, we can use more than one goroutine for a certain stage of the pipeline, or for a certain operation in our application.

We can create a small application that counts the occurrence of words which appear in a piece of text using such an approach. Let's create an initial producer stage that reads from a writer and returns a slice of words for that line:

```
func SourceLineWords(ctx context.Context, r io.ReadCloser) <-chan []string
{
    ch := make(chan []string)
    go func() {
        defer func() { r.Close(); close(ch) }()
        b := bytes.Buffer{}
        s := bufio.NewScanner(r)
        for s.Scan() {
            b.Reset()
            b.Write(s.Bytes())
            words := []string{}
            w := bufio.NewScanner(&b)
            w.Split(bufio.ScanWords)
            for w.Scan() {
                words = append(words, w.Text())
            }
            select {
```

```
                    case <-ctx.Done():
                        return
                    case ch <- words:
                    }
                }
            }()
        return ch
    }
```

We can now define another stage that will count the occurrence of these words. We will use this stage for the fan-out:

```
func WordOccurrence(ctx context.Context, src <-chan []string) <-chan
map[string]int {
    ch := make(chan map[string]int)
    go func() {
        defer close(ch)
        for v := range src {
            count := make(map[string]int)
            for _, s := range v {
                count[s]++
            }
            select {
            case <-ctx.Done():
                return
            case ch <- count:
            }
        }
    }()
    return ch
}
```

In order to use the first stage as a source for more than one instance of the second stage, we just need to create more than one counting stage with the same input channel:

```
ctx, canc := context.WithCancel(context.Background())
defer canc()
src := SourceLineWords(ctx,
    ioutil.NopCloser(strings.NewReader(cantoUno)))
count1, count2 := WordOccurrence(ctx, src), WordOccurrence(ctx, src)
```

Fan-in

Demuxing is a little bit more complicated because we don't need to receive data blindly in one goroutine or another – we actually need to synchronize a series of channels. A good approach to avoid race conditions is to create another channel where all the data from the various input channels will be received. We also need to make sure that this merge channel gets closed once all the channels are done. We also have to keep in mind that the channel will be closed if the context is cancelled. We are using `sync.Waitgroup` here to wait for all the channels to finish:

```
wg := sync.WaitGroup{}
merge := make(chan map[string]int)
wg.Add(len(src))
go func() {
    wg.Wait()
    close(merge)
}()
```

The problem is that we have two possible triggers for closing the channel: regular transmission ending and context cancelling.

We have to make sure that if the context ends, no message is sent to the outbound channel. Here, we are collecting the values from the input channels and sending them to the merge channel, but only if the context isn't complete. We do this in order to avoid a send operation being sent to a closed channel, which would make our application panic:

```
for _, ch := range src {
    go func(ch <-chan map[string]int) {
        defer wg.Done()
        for v := range ch {
            select {
            case <-ctx.Done():
                return
            case merge <- v:
            }
        }
    }(ch)
}
```

Finally, we can focus on the last operation, which uses the merge channel to execute our final word count:

```
count := make(map[string]int)
for {
    select {
    case <-ctx.Done():
```

```
        return count
case c, ok := <-merge:
    if !ok {
        return count
    }
    for k, v := range c {
        count[k] += v
    }
}
}
```

The application's `main` function, with the addition of the fan-in, will look as follows:

```
func main() {
    ctx, canc := context.WithCancel(context.Background())
    defer canc()
    src := SourceLineWords(ctx,
ioutil.NopCloser(strings.NewReader(cantoUno)))
    count1, count2 := WordOccurrence(ctx, src), WordOccurrence(ctx, src)
    final := MergeCounts(ctx, count1, count2)
    fmt.Println(final)
}
```

We can see that the fan-in is the most complex and critical part of the application. Let's recap the decisions that helped build a fan-in function that is free from panic or deadlock:

- Use a merge channel to collect values from the various input.
- Have `sync.WaitGroup` with a counter equal to the number of input channels.
- Use it in a separate goroutine and wait for it to close the channel.
- For each input channel, create a goroutine that transfers the values to the merge channel.
- Ensure that you send the record only if the context is not complete.
- Use the wait group's `done` function before exiting such a goroutine.

Following the preceding steps will allow us to use the merge channel with a simple `range`. In our example, we are also checking whether the context is complete before receiving from the channel in order to allow for an early exit from the goroutine.

Producers and consumers

Channels allow us to easily handle a scenario in which multiple consumers receive data from one producer and vice versa.

The case with a single producer and one consumer, as we have already seen, is pretty straightforward:

```
func main() {
    // one producer
    var ch = make(chan int)
    go func() {
        for i := 0; i < 100; i++ {
            ch <- i
        }
        close(ch)
    }()
    // one consumer
    var done = make(chan struct{})
    go func() {
        for i := range ch {
            fmt.Println(i)
        }
        close(done)
    }()
    <-done
}
```

The full example is available here: `https://play.golang.org/p/hNgehu62kjv`.

Multiple producers (N * 1)

Having multiple producers or consumers can be easily handled using wait groups. In the case of multiple producers, all the goroutines will share the same channel:

```
// three producer
var ch = make(chan string)
wg := sync.WaitGroup{}
wg.Add(3)
for i := 0; i < 3; i++ {
    go func(n int) {
        for i := 0; i < 100; i++ {
            ch <- fmt.Sprintln(n, i)
        }
        wg.Done()
    }(i)
}
go func() {
    wg.Wait()
    close(ch)
}()
```

The full example is available here: https://play.golang.org/p/4DqWKntl6sS.

They will use `sync.WaitGroup` to wait for each producer to finish before closing the channel.

Multiple consumers (1 * M)

The same reasoning applies with multiple consumers – they all receive from the same channel in different goroutines:

```go
func main() {
    // three consumers
    wg := sync.WaitGroup{}
    wg.Add(3)
    var ch = make(chan string)

    for i := 0; i < 3; i++ {
        go func(n int) {
            for i := range ch {
                fmt.Println(n, i)
            }
            wg.Done()
        }(i)
    }

    // one producer
    go func() {
        for i := 0; i < 10; i++ {
            ch <- fmt.Sprintln("prod-", i)
        }
        close(ch)
    }()

    wg.Wait()
}
```

The full example is available here: https://play.golang.org/p/_SWtw54ITFn.

In this case, `sync.WaitGroup` is used to wait for the application to end.

Multiple consumers and producers (N*M)

The last scenario is where we have an arbitrary number of producers (N) and another arbitrary number of consumers (M).

In this case, we need two waiting groups: one for the producer and another for the consumer:

```
const (
    N = 3
    M = 5
)
wg1 := sync.WaitGroup{}
wg1.Add(N)
wg2 := sync.WaitGroup{}
wg2.Add(M)
var ch = make(chan string)
```

This will be followed by a series of producers and consumers, each one in their own goroutine:

```
for i := 0; i < N; i++ {
    go func(n int) {
        for i := 0; i < 10; i++ {
            ch <- fmt.Sprintf("src-%d[%d]", n, i)
        }
        wg1.Done()
    }(i)
}

for i := 0; i < M; i++ {
    go func(n int) {
        for i := range ch {
            fmt.Printf("cons-%d, msg %q\n", n, i)
        }
        wg2.Done()
    }(i)
}
```

The final step is to wait for the WaitGroup producer to finish its work in order to close the channel.

Then, we can wait for the consumer channel to let all the messages be processed by the consumers:

```
wg1.Wait()
close(ch)
wg2.Wait()
```

Other patterns

So far, we've looked at the most common concurrency patterns that can be used. Now, we will focus on some that are less common but are worth mentioning.

Error groups

The power of `sync.WaitGroup` is that it allows us to wait for simultaneous goroutines to finish their jobs. We have already looked at how sharing context can allow us to give the goroutines an early exit if it's used correctly. The first concurrent operation, such as send or receive from a channel, is in the `select` block, together with the context completion channel:

```go
func main() {
    ctx, canc := context.WithTimeout(context.Background(), time.Second)
    defer canc()
    wg := sync.WaitGroup{}
    wg.Add(10)
    var ch = make(chan int)
    for i := 0; i < 10; i++ {
        go func(ctx context.Context, i int) {
            defer wg.Done()
            d := time.Duration(rand.Intn(2000)) * time.Millisecond
            time.Sleep(d)
            select {
            case <-ctx.Done():
                fmt.Println(i, "early exit after", d)
                return
            case ch <- i:
                fmt.Println(i, "normal exit after", d)
            }
        }(ctx, i)
    }
    go func() {
        wg.Wait()
        close(ch)
    }()
    for range ch {
    }
}
```

An improvement on this scenario is offered by the experimental `golang.org/x/sync/errgroup` package.

The built-in goroutines are always of the func() type, but this package allows us to execute func() error concurrently and return the first error that's received from the various goroutines.

This is very useful in scenarios where you launch more goroutines together and receive the first error. The errgroup.Group type can be used as a zero value, and its Do method takes func() error as an argument and launches the function concurrently.

The Wait method either waits for all the functions to finish successfully and returns nil, or it returns the first error that comes from any of the functions.

Let's create an example that defines a URL visitor, that is, a function that gets a URL string and returns func() error, which makes the call:

```
func visitor(url string) func() error {
    return func() (err error) {
        s := time.Now()
        defer func() {
            log.Println(url, time.Since(s), err)
        }()
        var resp *http.Response
        if resp, err = http.Get(url); err != nil {
            return
        }
        return resp.Body.Close()
    }
}
```

We can use it directly with the Go method and wait. This will return the error that was caused by the invalid URL:

```
func main() {
    eg := errgroup.Group{}
    var urlList = []string{
        "http://www.golang.org/",
        "http://invalidwebsite.hey/",
        "http://www.google.com/",
    }
    for _, url := range urlList {
        eg.Go(visitor(url))
    }
    if err := eg.Wait(); err != nil {
        log.Fatalln("Error:", err)
    }
}
```

The error group also allows us to create a group, along with a context, with the WithContext function. This context gets cancelled when the first error is received. The context's cancellation enables the Wait method to return right away, but it also allows an early exit in the goroutines in your functions.

We can create a similar func() error creator that will send values into a channel until the context is closed. We will introduce a small chance (1%) of raising an error:

```
func sender(ctx context.Context, ch chan<- string, n int) func() error {
    return func() (err error) {
        for i := 0; ; i++ {
            if rand.Intn(100) == 42 {
                return errors.New("the answer")
            }
            select {
            case ch <- fmt.Sprintf("[%d]%d", n, i):
            case <-ctx.Done():
                return nil
            }
        }
    }
}
```

We will generate an error group and a context with the dedicated function and use it to launch several instances of the function. We will receive this in a separate goroutine while we wait for the group. After the wait is over, we will make sure that there are no more values being sent to the channel (this would cause a panic) by waiting an extra second:

```
func main() {
    eg, ctx := errgroup.WithContext(context.Background())
    ch := make(chan string)
    for i := 0; i < 10; i++ {
        eg.Go(sender(ctx, ch, i))
    }
    go func() {
        for s := range ch {
            log.Println(s)
        }
    }()
    if err := eg.Wait(); err != nil {
        log.Println("Error:", err)
    }
    close(ch)
    log.Println("waiting...")
    time.Sleep(time.Second)
}
```

As expected, thanks to the `select` statement within the context, the application runs seamlessly and does not panic.

Leaky bucket

We saw how to build a rate limiter using ticker in the previous chapters: by using `time.Ticker` to force a client to await its turn in order to get served. There is another take on rate limiting of services and libraries that's known as the **leaky bucket**. The name evokes an image of a bucket with a few holes in it. If you are filling it, you have to be careful to not put too much water into the bucket, otherwise it's going to overflow. Before adding more water, you need to wait for the level to drop – the speed at which this happens will depend on the size of the bucket and the number of the holes it has. We can easily understand what this concurrency pattern does by taking a look at the following analogy:

- The water going through the holes represents requests that have been completed.
- The water that's overflowing from the bucket represents the requests that have been discarded.

The bucket will be defined by two attributes:

- **Rate**: The ideal amount of requests per time if the frequency of requests is lower.
- **Capacity**: The number of requests that can be done at the same time before the resource turns unresponsive temporarily.

The bucket has a maximum capacity, so when requests are made with a frequency higher than the rate specified, this capacity starts dropping, just like when you're putting too much water in and the bucket starts to overflow. If the frequency is zero or lower than the rate, the bucket will slowly gain its capacity, and so the water will be slowly drained.

The data structure of the leaky bucket will have a capacity and a counter for the requests that are available. This counter will be the same as the capacity on creation, and will drop each time requests are executed. The rate specifies how often the status needs to be reset to the capacity:

```
type bucket struct {
    capacity uint64
    status uint64
}
```

When creating a new bucket, we should also take care of the status reset. We can use a goroutine for this and use a context to terminate it correctly. We can create a ticker using the rate and then use these ticks to reset the status. We need to use the atomic package to ensure it is thread-safe:

```go
func newBucket(ctx context.Context, cap uint64, rate time.Duration) *bucket
{
    b := bucket{capacity: cap, status: cap}
    go func() {
        t := time.NewTicker(rate)
        for {
            select {
            case <-t.C:
                atomic.StoreUint64(&b.status, b.capacity)
            case <-ctx.Done():
                t.Stop()
                return
            }
        }
    }()
    return &b
}
```

When we're adding to the bucket, we can check the status and act accordingly:

- If the status is 0, we cannot add anything.
- If the amount to add is higher than the availability, we add what we can.
- We add the full amount otherwise:

```go
func (b *bucket) Add(n uint64) uint64 {
    for {
        r := atomic.LoadUint64(&b.status)
        if r == 0 {
            return 0
        }
        if n > r {
            n = r
        }
        if !atomic.CompareAndSwapUint64(&b.status, r, r-n) {
            continue
        }
        return n
    }
}
```

We are using a loop to try atomic swap operations until they succeed to ensure that what we get with the Load operation doesn't change when we are doing a **compare and swap** (**CAS**).

The bucket can be used in a client that will try to add a random amount to the bucket and will log its result:

```
type client struct {
    name string
    max int
    b *bucket
    sleep time.Duration
}

func (c client) Run(ctx context.Context, start time.Time) {
    for {
        select {
        case <-ctx.Done():
            return
        default:
            n := 1 + rand.Intn(c.max-1)
            time.Sleep(c.sleep)
            e := time.Since(start).Seconds()
            a := c.b.Add(uint64(n))
            log.Printf("%s tries to take %d after %.02fs, takes
                %d", c.name, n, e, a)
        }
    }
}
```

We can use more clients concurrently so that having concurrent access to resources will have the following result:

- Some goroutines will be adding what they expect to the bucket.
- One goroutine will finally fill the bucket by adding a quantity that is equal to the remaining capacity, even if the amount that they are trying to add is higher.
- The other goroutines will not be able to add to the bucket until the capacity is reset:

```
func main() {
    ctx, canc := context.WithTimeout(context.Background(),
time.Second)
    defer canc()
    start := time.Now()
    b := newBucket(ctx, 10, time.Second/5)
    t := time.Second / 10
```

```
        for i := 0; i < 5; i++ {
            c := client{
                name: fmt.Sprint(i),
                b: b,
                sleep: t,
                max: 5,
            }
            go c.Run(ctx, start)
        }
        <-ctx.Done()
    }
```

Sequencing

In concurrent scenarios with multiple goroutines, we may need to have a synchronization between goroutines, such as in a scenario where each goroutine needs to wait for its turn after sending.

A use case for this scenario could be a turn-based application wherein different goroutines are sending messages to the same channel, and each one of them has to wait until all the others have finished before they can send it again.

A very simple implementation of this scenario can be obtained using private channels between the main goroutine and the senders. We can define a very simple structure that carries both messages and a Wait channel. It will have two methods – one for marking the transaction as done and another one that waits for such a signal – when it uses a channel underneath. The following method shows this:

```
type msg struct {
    value string
    done chan struct{}
}

func (m *msg) Wait() {
    <-m.done
}

func (m *msg) Done() {
    m.done <- struct{}{}
}
```

We can create a source of messages with a generator. We can use a random delay with the
send operation. After each send, we wait for the signal that is obtained by calling
the Done method. We always use context to keep everything free from leaks:

```
func send(ctx context.Context, v string) <-chan msg {
    ch := make(chan msg)
    go func() {
        done := make(chan struct{})
        for i := 0; ; i++ {
            time.Sleep(time.Duration(float64(time.Second/2) *
rand.Float64()))
            m := msg{fmt.Sprintf("%s msg-%d", v, i), done}
            select {
            case <-ctx.Done():
                close(ch)
                return
            case ch <- m:
                m.Wait()
            }
        }
    }()
    return ch
}
```

We can use a fan-in to put all of the channels into one, singular channel:

```
func merge(ctx context.Context, sources ...<-chan msg) <-chan msg {
    ch := make(chan msg)
    go func() {
        <-ctx.Done()
        close(ch)
    }()
    for i := range sources {
        go func(i int) {
            for {
                select {
                case v := <-sources[i]:
                    select {
                    case <-ctx.Done():
                        return
                    case ch <- v:
                    }
                }
            }
        }(i)
    }
    return ch
}
```

The main application will be receiving from the merged channel until it's closed. When it receives one message from each channel, the channel will be blocked, waiting for the Done method signal to be called by the main goroutine.

This specific configuration will allow the main goroutine to receive just one message from each channel. When the message count reaches the number of goroutines, we can call Done from the main goroutine and reset the list so that the other goroutines will be unlocked and be able to send messages again:

```go
func main() {
    ctx, canc := context.WithTimeout(context.Background(), time.Second)
    defer canc()
    sources := make([]<-chan msg, 5)
    for i := range sources {
        sources[i] = send(ctx, fmt.Sprint("src-", i))
    }
    msgs := make([]msg, 0, len(sources))
    start := time.Now()
    for v := range merge(ctx, sources...) {
        msgs = append(msgs, v)
        log.Println(v.value, time.Since(start))
        if len(msgs) == len(sources) {
            log.Println("*** done ***")
            for _, m := range msgs {
                m.Done()
            }
            msgs = msgs[:0]
            start = time.Now()
        }
    }
}
```

Running the application will result in all the goroutines sending a message to the main one once. Each of them will be waiting for everyone to send their message. Then, they will start sending messages again. This results in messages being sent in rounds, as expected.

Summary

In this chapter, we looked at some specific concurrency patterns for our applications. We learned that generators are functions that return channels, and also feed such channels with data and close them when there is no more data. We also saw that we can use a context to allow the generator to exit early.

Next, we focused on pipelines, which are stages of execution that use channels for communication. They can either be source, which doesn't require any input; destination, which doesn't return a channel; or intermediate, which receives a channel as input and returns one as output.

Another pattern is the multiplexing and demultiplexing one, which consists of spreading a channel to different goroutines and combining several channels into one. It is often referred to as *fan-out fan-in*, and it allows us to execute different operations concurrently on a set of data.

Finally, we learned how to implement a better version of the rate limiter called **leaky bucket**, which limits the number of requests in a specific amount of time. We also looked at the sequencing pattern, which uses a private channel to signal to all of the sending goroutines when they are allowed to send data again.

In this next chapter, we are going to introduce the first of two extra topics that were presented in the *Sequencing* section. It is here that we will demonstrate how to use reflection to build generic code that adapts to any user-provided type.

Questions

1. What is a generator? What are its responsibilities?
2. How could you describe a pipeline?
3. What type of stage gets a channel and returns one?
4. What is the difference between fan-in and fan-out?

5

Section 5: A Guide to Using Reflection and CGO

This section focuses on two tools that are very controversial—reflection, which allows the creation of generic code but has a great cost in terms of performance, and CGO, which allows the use of C code in a Go application but makes it more complex to debug and control the application.

This section consists of the following chapters:

- Chapter 15, *Using Reflection*
- Chapter 16, *Using CGO*

15
Using Reflection

This chapter is about **reflection**, a tool that allows an application to inspect its own code, overcoming some of the limitations imposed by Go static typing and its lack of generics. This can be very helpful, for instance, for producing packages that are capable of handling any type of input that they receive.

The following topics will be covered in this chapter:

- Understanding interfaces and type assertions
- Learning about interaction with basic types
- Using reflection with complex types
- Evaluating the cost of reflection
- Learning the best practices of reflection usage

Technical requirements

This chapter requires Go to be installed and your favorite editor to be set up. For more information, refer to Chapter 3, *An Overview of Go*.

What's reflection?

Reflection is a very powerful feature that allows **meta-programming**, the capability of an application to examine its own structure. It's very useful to analyze the types in an application at runtime, and it is used in many encoding packages such as JSON and XML.

Type assertions

We briefly mentioned how type assertions work in `Chapter 3`, *An Overview of Go*. A type assertion is an operation that allows us to go from interface to concrete type and vice versa. It takes the following form:

```
# unsafe assertion
v := SomeVar.(SomeType)

# safe assertion
v, ok := SomeVar.(SomeType)
```

The first version is unsafe, and it assigns a value to a single variable.

Using assertion as an argument of a function also counts as unsafe. This type of operation panics if the assertion is wrong:

```
func main() {
    var a interface{} = "hello"
    fmt.Println(a.(string)) // ok
    fmt.Println(a.(int))    // panics!
}
```

A full example is available here: `https://play.golang.org/p/hNN87SuprGR`.

The second version uses a Boolean a as second value, and it will show the success of the operation. If the assertion is not possible, the first value will always be a zero value for the asserted type:

```
func main() {
    var a interface{} = "hello"
    s, ok := a.(string) // true
    fmt.Println(s, ok)
    i, ok := a.(int) // false
    fmt.Println(i, ok)
}
```

A full example is available here: `https://play.golang.org/p/BIba2ywkNF_j`.

Interface assertion

Assertion can also be done from one interface to another. Imagine having two different interfaces:

```
type Fooer interface {
    Foo()
}

type Barer interface {
    Bar()
}
```

Let's define a type that implements one of them and another that implements both:

```
type A int

func (A) Foo() {}

type B int

func (B) Bar() {}
func (B) Foo() {}
```

If we define a new variable for the first interface, the assertion to the second is going to be successful only if the underlying value has a type that implements both; otherwise, it's going to fail:

```
func main() {
    var a Fooer
    a = A(0)
    v, ok := a.(Barer)
    fmt.Println(v, ok)

    a = B(0)
    v, ok = a.(Barer)
    fmt.Println(v, ok)
}
```

A full example is available here: https://play.golang.org/p/bX2rnw5pRXJ.

A use case scenario could be having the `io.Reader` interface, checking out whether it is also an `io.Closer` interface, and wrapping it in the `ioutil.NopCloser` function (which returns an `io.ReadCloser` interface) if not:

```
func Closer(r io.Reader) io.ReadCloser {
    if rc, ok := r.(io.ReadCloser); ok {
        return rc
    }
    return ioutil.NopCloser(r)
}

func main() {
    log.Printf("%T", Closer(&bytes.Buffer{}))
    log.Printf("%T", Closer(&os.File{}))
}
```

A full example is available here: `https://play.golang.org/p/hUEsDYHFE7i`.

There is an important aspect to interfaces that we need to underline before jumping onto reflection—its representation is always a tuple interface-value where the value is a concrete type and cannot be another interface.

Understanding basic mechanics

The `reflection` package allows you to extract the type and value from any `interface{}` variable. This can be done using the following:

- Using `reflection.TypeOf` returns the type of the interface in a `reflection.Type` variable.
- The `reflection.ValueOf` function returns the value of the interface using the `reflection.Value` variable.

Value and Type methods

A `reflect.Value` type also carries information of the type that can be retrieved with the `Type` method:

```
func main() {
    var a interface{} = int64(23)
    fmt.Println(reflect.TypeOf(a).String())
    // int64
    fmt.Println(reflect.ValueOf(a).String())
```

```
    // <int64 Value>
    fmt.Println(reflect.ValueOf(a).Type().String())
    // int64
}
```

A full example is available here: `https://play.golang.org/p/tmYuMc4AF1T`.

Kind

Another important property of `reflect.Type` is `Kind`, which is an enumeration of basic types and generic complex types. The main relationship between `reflect.Kind` and `reflect.Type` is that the first represents the memory representation of the second.

For built-in types, `Kind` and `Type` are the same, but for custom types they will differ—the `Type` value will be what is expected, but the `Kind` value will be one of the built-in ones on which the custom type is defined:

```
func main() {
    var a interface{}

    a = "" // built in string
    t := reflect.TypeOf(a)
    fmt.Println(t.String(), t.Kind())

    type A string // custom type
    a = A("")
    t = reflect.TypeOf(a)
    fmt.Println(t.String(), t.Kind())
}
```

A full example is available here: `https://play.golang.org/p/qjiouk88INn`.

For the composite type, it will reflect just the main type and not the underlying ones. This means that a pointer to a structure or to an integer is the same kind, `reflect.Pointer`:

```
func main() {
    var a interface{}

    a = new(int) // int pointer
    t := reflect.TypeOf(a)
    fmt.Println(t.String(), t.Kind())

    a = new(struct{}) // struct pointer
    t = reflect.TypeOf(a)
    fmt.Println(t.String(), t.Kind())
}
```

A full example is available here: `https://play.golang.org/p/-uJjZvTuzVf`.

The same reasoning applies to all the other composite types, such as arrays, slices, maps, and channels.

Value to interface

In the same way that we can get `reflect.Value` from any `interface{}` value, we can execute the reverse operation and obtain `interface{}` from `reflect.Value`. This is done using the `Interface` method of the reflected value, and be cast to a concrete type if necessary. If the interested method or function accepts an empty interface, such as `json.Marshal` or `fmt.Println`, the returned value can be passed directly without any casting:

```
func main() {
    var a interface{} = int(12)
    v := reflect.ValueOf(a)
    fmt.Println(v.String())
    fmt.Printf("%v", v.Interface())
}
```

A full example is available here: `https://play.golang.org/p/1942Dhm5sap`.

Manipulating values

Transforming values in their reflection and going back to the value is not very useful if the values themselves cannot be changed. That's why our next step is seeing how to change them using the `reflection` package.

Changing values

There is a series of methods of the `reflect.Value` type that allow you to change the underlying value:

- `Set`: Uses another `reflect.Value`
- `SetBool`: Boolean
- `SetBytes`: Byte slice
- `SetComplex`: Any complex type
- `SetFloat`: Any float type

- `SetInt`: Any signed integer type
- `SetPointer`: A pointer
- `SetString`: A string
- `SetUint`: Any unsigned integer

In order to set a value, it needs to editable, and this happens in specific conditions. To verify this, there is a method, `CanSet`, which returns `true` if a value can be changed. If the value cannot be changed and a `Set` method is called anyway, the application will panic:

```
func main() {
    var a = int64(12)
    v := reflect.ValueOf(a)
    fmt.Println(v.String(), v.CanSet())
    v.SetInt(24)
}
```

A full example is available here: `https://play.golang.org/p/hKn8qNtn0gN`.

In order to be changed, a value needs to be addressable. A value is addressable if it's possible to modify the actual storage where the object is saved. When creating a new value using a basic built-in type, such as `string` , what gets passed to the function is `interface{}` , which hosts a copy of the string.

Changing this copy would result in a variation of the copy with no effect on the original variable. This would be incredibly confusing, and it would make the usage of a sensible tool such as reflection, even harder. That's why, instead of this useless behavior, the `reflect` package panics—it's a design choice. This explains why the last example panicked.

We can create `reflect.Value` using the pointer to the value we want to change, and access the value using the `Elem` method. This will give us a value that is addressable because we copied the pointer instead of the value, so the reflected value is still a pointer to the variable:

```
func main() {
    var a = int64(12)
    v := reflect.ValueOf(&a)
    fmt.Println(v.String(), v.CanSet())
    e := v.Elem()
    fmt.Println(e.String(), e.CanSet())
    e.SetInt(24)
    fmt.Println(a)
}
```

A full example is available here: `https://play.golang.org/p/-X5JsBrlr4Q`.

Creating new values

The `reflect` package allows us also to create new values using types. There are several functions that allow us to create a value:

- `MakeChan` creates a new channel value
- `MakeFunc` creates a new function value
- `MakeMap` and `MakeMapWithSize` creates a new map value
- `MakeSlice` creates a new slice value
- `New` creates a new pointer to the type
- `NewAt` creates a new pointer to the type using the selected address
- `Zero` creates a zero value of the selected type

The following code shows how to create new values in a couple of different ways:

```go
func main() {
    t := reflect.TypeOf(int64(100))
    // zero value
    fmt.Printf("%#v\n", reflect.Zero(t))
    // pointer to int
    fmt.Printf("%#v\n", reflect.New(t))
}
```

A full example is available here: `https://play.golang.org/p/wCTILSK1F1C`.

Handling complex types

After seeing how to handle the reflection basics, we will now see how complex data types such as structures and maps can also be handled using reflection.

Data structures

For changeability, structures work in exactly the same way as the basic types; we need to obtain the reflection of the pointer, then access its element in order to be able to change the value, because using the structure directly would produce a copy of it and it would panic when changing values.

We can replace the value of the entire structure using the `Set` method, after obtaining the new value's reflection:

```go
func main() {
    type X struct {
        A, B int
        c string
    }
    var a = X{10, 100, "apple"}
    fmt.Println(a)
    e := reflect.ValueOf(&a).Elem()
    fmt.Println(e.String(), e.CanSet())
    e.Set(reflect.ValueOf(X{1, 2, "banana"}))
    fmt.Println(a)
}
```

A full example is available here: `https://play.golang.org/p/mjb3gJw5CeA`.

Changing fields

Individual fields can also be modified using `Field` methods:

- `Field` returns a field using its index
- `FieldByIndex` returns a nested field using a series of indexes
- `FieldByName` returns a field using its name
- `FieldByNameFunc` returns a field using `func(string) bool` in the name

Let's define a structure to change the values of the fields, using both simple and complex types, with at least one unexported field:

```go
type A struct {
    B
    x int
    Y int
    Z int
}

type B struct {
    F string
    G string
}
```

Now that we have the structure, we can try to access the fields in different ways:

```
func main() {
    var a A
    v := reflect.ValueOf(&a)
    func() {
        // trying to get fields from ptr panics
        defer func() {
            log.Println("panic:", recover())
        }()
        log.Printf("%s", v.Field(1).String())
    }()
    v = v.Elem()
    // changing fields by index
    for i := 0; i < 4; i++ {
        f := v.Field(i)
        if f.CanSet() && f.Type().Kind() == reflect.Int {
            f.SetInt(42)
        }
    }
    // changing nested fields by index
    v.FieldByIndex([]int{0, 1}).SetString("banana")

    // getting fields by name
    v.FieldByName("B").FieldByName("F").SetString("apple")

    log.Printf("%+v", a)
}
```

A full example is available here: `https://play.golang.org/p/z5slFkIU5UE`.

When working with `reflect.Value` and structure fields, what you get are other values, indistinguishable from the struct. When handling `reflect.Type` instead, you obtain a `reflect.StructField` structure, which is another type that carries all the information of the field with it.

Using tags

A structure field carries plenty of information, from the field name and index to its tag:

```
type StructField struct {
    Name string
    PkgPath string

    Type Type      // field type
    Tag StructTag  // field tag string
```

```
    Offset uintptr // offset within struct, in bytes
    Index []int    // index sequence for Type.FieldByIndex
    Anonymous bool // is an embedded field
}
```

A `reflect.StructField` value can be obtained using the `reflect.Type` methods:

- `Field`
- `FieldByName`
- `FieldByIndex`

They are the same methods used by `reflect.Value`, but they return different types. The `NumField` method returns the total number of fields for the structure, allowing us to execute an iteration:

```
type Person struct {
    Name string `json:"name,omitempty" xml:"-"`
    Surname string `json:"surname,omitempty" xml:"-"`
}

func main() {
    v := reflect.ValueOf(Person{"Micheal", "Scott"})
    t := v.Type()
    fmt.Println("Type:", t)
    for i := 0; i < t.NumField(); i++ {
        fmt.Printf("%v: %v\n", t.Field(i).Name, v.Field(i))
    }
}
```

A full example is available here: `https://play.golang.org/p/nkEADg77zFC`.

Tags are really central to reflection because they can store extra information about a field and how other packages behave with it. To add a tag to a field, it needs to be inserted after the field name and type in a string, which should have a `key:"value"` structure. One field can have multiple tuples in its tag, and each pair is separated by a space. Let's look at a practical example:

```
type A struct {
    Name    string `json:"name,omitempty" xml:"-"`
    Surname string `json:"surname,omitempty" xml:"-"`
}
```

This structure has two fields, both with tags, and each tag has two pairs. The Get method returns the value for a specific key:

```go
func main() {
    t := reflect.TypeOf(A{})
    fmt.Println(t)
    for i := 0; i < t.NumField(); i++ {
        f := t.Field(i)
        fmt.Printf("%s JSON=%s XML=%s\n", f.Name, f.Tag.Get("json"),
f.Tag.Get("xml"))
    }
}
```

A full example is available here: https://play.golang.org/p/P-Te8O1Hyyn.

Maps and slices

You can easily use reflection to read and manipulate maps and slices. Since they are such important tools for writing applications, let's see how to execute an operation using reflection.

Maps

A map type allows you to get the type of both value and key, using the Key and Elem methods:

```go
func main() {
    maps := []interface{}{
        make(map[string]struct{}),
        make(map[int]rune),
        make(map[float64][]byte),
        make(map[int32]chan bool),
        make(map[[2]string]interface{}),
    }
    for _, m := range maps {
        t := reflect.TypeOf(m)
        fmt.Printf("%s k:%-10s v:%-10s\n", m, t.Key(), t.Elem())
    }
}
```

A full example is available here: `https://play.golang.org/p/j__1jtgy-56`.

The values can be accessed in all the ways that a map can be accessed normally:

- By getting a value using a key
- By ranging over keys
- By ranging over values

Let's see how it works in a practical example:

```go
func main() {
    m := map[string]int64{
        "a": 10,
        "b": 20,
        "c": 100,
        "d": 42,
    }

    v := reflect.ValueOf(m)

    // access one field
    fmt.Println("a", v.MapIndex(reflect.ValueOf("a")))
    fmt.Println()

    // range keys
    for _, k := range v.MapKeys() {
        fmt.Println(k, v.MapIndex(k))
    }
    fmt.Println()

    // range keys and values
    i := v.MapRange()
    for i.Next() {
        fmt.Println(i.Key(), i.Value())
    }
}
```

Note that we don't need to pass a pointer to the map to make it addressable, because maps are already pointers.

Each method is pretty straightforward and depends on the type of access you need to the map. Setting values is also possible, and should always be possible because maps are passed by reference. The following snippet shows a practical example:

```go
func main() {
    m := map[string]int64{}
    v := reflect.ValueOf(m)
```

```
        // setting one field
        v.SetMapIndex(reflect.ValueOf("key"), reflect.ValueOf(int64(1000)))

        fmt.Println(m)
}
```

A full example is available here: `https://play.golang.org/p/JxK_8VPoWU0`.

It is also possible to use this method to unset a variable, like we do when calling the `delete` function using the zero value of `reflect.Value` as a second argument:

```
func main() {
        m := map[string]int64{"a": 10}
        fmt.Println(m, len(m))
        v := reflect.ValueOf(m)

        // deleting field
        v.SetMapIndex(reflect.ValueOf("a"), reflect.Value{})

        fmt.Println(m, len(m))
}
```

A full example is available here: `https://play.golang.org/p/4bPqfmaKzTC`.

The output will have one less field as it gets deleted because the length of the map decreases after `SetMapIndex`.

Slices

A slice allows you to get its size with the `Len` method and to access its elements using the `Index` method. Let's see that in action in the following code:

```
func main() {
        m := []int{10, 20, 100}
        v := reflect.ValueOf(m)

        for i := 0; i < v.Len(); i++ {
                fmt.Println(i, v.Index(i))
        }
}
```

A full example is available here: `https://play.golang.org/p/ifq0O6bFIZc`.

Since it is always possible to get the address of a slice element, it is also possible to use `reflect.Value` to change the content of the respective element in the slice:

```
func main() {
    m := []int64{10, 20, 100}
    v := reflect.ValueOf(m)

    for i := 0; i < v.Len(); i++ {
        v.Index(i).SetInt(v.Index(i).Interface().(int64) * 2)
    }
    fmt.Println(m)
}
```

A full example is available here: `https://play.golang.org/p/onuIvWyQ7GY`.

It is also possible to append to a slice using the `reflect` package. If the value is obtained from the pointer to the slice, the result of this operation can also be used to replace the original slice:

```
func main() {
    var s = []int{1, 2}
    fmt.Println(s)

    v := reflect.ValueOf(s)
    // same as append(s, 3)
    v2 := reflect.Append(v, reflect.ValueOf(3))
    // s can't and does not change
    fmt.Println(v.CanSet(), v, v2)
    // using the pointer allows change
    v = reflect.ValueOf(&s).Elem()
    v.Set(v2)
    fmt.Println(v.CanSet(), v, v2)
}
```

A full example is available here: `https://play.golang.org/p/2hXRg7Ih9wk`.

Functions

Method and function handling with reflection allow you to gather information about the signature of a certain entry and also to invoke it.

Analyzing a function

There are a few methods of `reflect.Type` in the package that will return information about a function. These methods are as follows:

- `NumIn`: Returns the number of input arguments of the function
- `In`: Returns the selected input argument
- `IsVariadic`: Tells you if the last argument of the function is variadic
- `NumOut`: Returns the number of output values returned by the function
- `Out`: Returns the `Type` value of the select output

Note that all these methods will panic if the kind of `reflect.Type` is not `Func`. We can test these methods by defining a series of functions:

```go
func Foo() {}

func Bar(a int, b string) {}

func Baz(a int, b string) (int, error) { return 0, nil }

func Qux(a int, b ...string) (int, error) { return 0, nil }
```

Now we can use the method from `reflect.Type` to obtain information about them:

```go
func main() {
    for _, f := range []interface{}{Foo, Bar, Baz, Qux} {
        t := reflect.TypeOf(f)
        name := runtime.FuncForPC(reflect.ValueOf(f).Pointer()).Name()
        in := make([]reflect.Type, t.NumIn())
        for i := range in {
            in[i] = t.In(i)
        }
        out := make([]reflect.Type, t.NumOut())
        for i := range out {
            out[i] = t.Out(i)
        }
        fmt.Printf("%q %v %v %v\n", name, in, out, t.IsVariadic())
    }
}
```

A full example is available here: `https://play.golang.org/p/LAjjhw8Et60`.

In order to obtain the name of the functions, we use the `runtime.FuncForPC` function, which returns `runtime.Func` containing methods that will expose runtime information about the function—name, `file`, and `line`. The function takes `uintptr` as an argument, which can be obtained with `reflect.Value` of the function and its `Pointer` method.

Invoking a function

While the type of the function shows information about it, in order to call a function, we need to use its value.

We will pass the function a list of argument values and get back the ones returned by the function call:

```go
func main() {
    for _, f := range []interface{}{Foo, Bar, Baz, Qux} {
        v, t := reflect.ValueOf(f), reflect.TypeOf(f)
        name := runtime.FuncForPC(v.Pointer()).Name()
        in := make([]reflect.Value, t.NumIn())
        for i := range in {
            switch a := t.In(i); a.Kind() {
            case reflect.Int:
                in[i] = reflect.ValueOf(42)
            case reflect.String:
                in[i] = reflect.ValueOf("42")
            case reflect.Slice:
                switch a.Elem().Kind() {
                case reflect.Int:
                    in[i] = reflect.ValueOf(21)
                case reflect.String:
                    in[i] = reflect.ValueOf("21")
                }
            }
        }
        out := v.Call(in)
        fmt.Printf("%q %v%v\n", name, in, out)
    }
}
```

A full example is available here: `https://play.golang.org/p/jPxO_G7YP2I`.

Channels

Reflection allows us to create channels, send and receive data, and also to use `select` statements.

Creating channels

A new channel can be created via the `reflect.MakeChan` function, which requires a `reflect.Type` interface value and a size:

```
func main() {
    t := reflect.ChanOf(reflect.BothDir, reflect.TypeOf(""))
    v := reflect.MakeChan(t, 0)
    fmt.Printf("%T\n", v.Interface())
}
```

A full example is available here: https://play.golang.org/p/7_RLtzjuTcz.

Sending, receiving, and closing

The `reflect.Value` type offers a few methods that have to be used exclusively with channels, `Send` and `Recv` for sending and receiving, and `Close` for closing channels. Let's take a look at a sample use case of these functions and methods:

```
func main() {
    t := reflect.ChanOf(reflect.BothDir, reflect.TypeOf(""))
    v := reflect.MakeChan(t, 0)
    go func() {
        for i := 0; i < 10; i++ {
            v.Send(reflect.ValueOf(fmt.Sprintf("msg-%d", i)))
        }
        v.Close()
    }()
    for msg, ok := v.Recv(); ok; msg, ok = v.Recv() {
        fmt.Println(msg)
    }
}
```

A full example is available here: https://play.golang.org/p/Gp8JJmDbLIL.

Select statement

A `select` statement can be executed with the `reflect.Select` function. Each case is represented by a data structure:

```
type SelectCase struct {
    Dir  SelectDir // direction of case
    Chan Value      // channel to use (for send or receive)
    Send Value      // value to send (for send)
}
```

It contains the direction of the operation and both the channel and the value (for send operations). The direction can be either send, receive, or none (for default statements):

```
func main() {
    v := reflect.ValueOf(make(chan string, 1))
    fmt.Println("sending", v.TrySend(reflect.ValueOf("message"))) // true 1
1
    branches := []reflect.SelectCase{
        {Dir: reflect.SelectRecv, Chan: v, Send: reflect.Value{}},
        {Dir: reflect.SelectSend, Chan: v, Send: reflect.ValueOf("send")},
        {Dir: reflect.SelectDefault},
    }

    // send, receive and default
    i, recv, closed := reflect.Select(branches)
    fmt.Println("select", i, recv, closed)

    v.Close()
    // just default and receive
    i, _, closed = reflect.Select(branches[:2])
    fmt.Println("select", i, closed) // 1 false
}
```

A full example is available here: `https://play.golang.org/p/_DgSYRIBkJA`.

Reflecting on reflection

After talking how about how reflection works in all its aspects, we will now focus on its downside, when it is used in the standard library, and when to use it in packages.

Performance cost

Reflection allows code to be flexible and handles unknown data types by analyzing their memory representation. This is not cost-free and, besides complexity, another aspect that reflection influences is performance.

We can create a couple of examples to demonstrate how some trivial operations are much slower using reflection. We can create a timeout and keep repeating these operations in goroutines. Both routines will terminate when the timeout expires, and we will compare the results:

```go
func baseTest(fn1, fn2 func(int)) {
    ctx, canc := context.WithTimeout(context.Background(), time.Second)
    defer canc()
    go func() {
        for i := 0; ; i++ {
            select {
            case <-ctx.Done():
                return
            default:
                fn1(i)
            }
        }
    }()
    go func() {
        for i := 0; ; i++ {
            select {
            case <-ctx.Done():
                return
            default:
                fn2(i)
            }
        }
    }()
    <-ctx.Done()
}
```

We can compare normal map writing with the same operation done with reflection:

```
func testMap() {
    m1, m2 := make(map[int]int), make(map[int]int)
    m := reflect.ValueOf(m2)
    baseTest(func(i int) { m1[i] = i }, func(i int) {
        v := reflect.ValueOf(i)
        m.SetMapIndex(v, v)
    })
    fmt.Printf("normal %d\n", len(m1))
    fmt.Printf("reflect %d\n", len(m2))
}
```

We can also test out how fast the reading is and the settings of a structure field, with and without reflection:

```
func testStruct() {
    type T struct {
        Field int
    }
    var m1, m2 T
    m := reflect.ValueOf(&m2).Elem()
    baseTest(func(i int) { m1.Field++ }, func(i int) {
        f := m.Field(0)
        f.SetInt(int64(f.Interface().(int) + 1))
    })
    fmt.Printf("normal %d\n", m1.Field)
    fmt.Printf("reflect %d\n", m2.Field)
}
```

There is at least a 50% performance drop when executing operation via reflection, compared to the standard, static way of doing things. This drop could be very critical when performance is a very important priority in an application, but if that's not the case, then the use of reflection could be a reasonable call.

Usage in the standard library

There are many different packages in the standard library that use the reflect package:

- archive/tar
- context
- database/sql
- encoding/asn1
- encoding/binary

- encoding/gob
- encoding/json
- encoding/xml
- fmt
- html/template
- net/http
- net/rpc
- sort/slice
- text/template

We can reason about their approach to reflection, taking as an example the encoding packages. Each of these packages offers interfaces for encoding and decoding, for instance, the encoding/json package. We have the following interfaces defined:

```
type Marshaler interface {
    MarshalJSON() ([]byte, error)
}

type Unmarshaler interface {
    UnmarshalJSON([]byte) error
}
```

The package first looks if the unknown type implements the interface while decoding or encoding, and, if not, it uses reflection. We can think of reflection as a last resource that the package uses. Even the sort package has a generic slice method that takes any slice using reflection to set values and a sorting interface that avoids using reflection.

There are other packages, such as text/template and html/template, that read runtime text files with instructions on which method or field to access or to use. In this case, there is no other way than reflection to accomplish it, and there is no interface that can avoid it.

Using reflection in a package

After seeing how reflection works and the kind of complications that it adds to code, we can think about using it in a package we are writing. One of the Go proverbs, from its creator Rob Pike, comes to the rescue:

Clear is better than clever. Reflection is never clear.

The power of reflection is huge, but it also comes at the expense of making code more complicated and implicit. It should be used only when it's extremely necessary, as in the template scenario, and should be avoided in any other case, or at least offer an interface to avoid it, as in the encoding packages.

Property files

We can try to use reflection to create a package that reads property files.

We can use reflection to create a package that reads property files:

1. The first thing we should do is define an interface that avoids using reflection:

```
type Unmarshaller interface {
    UnmarshalProp([]byte) error
}
```

2. Then, we can define a decoder structure that will feed on an io.Reader instance, using a line scanner to read the individual properties:

```
type Decoder struct {
    scanner *bufio.Scanner
}

func NewDecoder(r io.Reader) *Decoder {
    return &Decoder{scanner: bufio.NewScanner(r)}
}
```

3. The decoder will also be used by the Unmarshal method:

```
func Unmarshal(data []byte, v interface{}) error {
    return NewDecoder(bytes.NewReader(data)).Decode(v)
}
```

4. We can reduce the number of uses of reflection that we will do by building a cache of field names and indices. This will be helpful because the value of a field in reflection can only be accessed by an index, and not by a name:

```
var cache = make(map[reflect.Type]map[string]int)

func findIndex(t reflect.Type, k string) (int, bool) {
    if v, ok := cache[t]; ok {
        n, ok := v[k]
        return n, ok
```

```
    }
    m := make(map[string]int)
    for i := 0; i < t.NumField(); i++ {
        f := t.Field(i)
        if s := f.Name[:1]; strings.ToLower(s) == s {
            continue
        }
        name := strings.ToLower(f.Name)
        if tag := f.Tag.Get("prop"); tag != "" {
            name = tag
        }
        m[name] = i
    }
    cache[t] = m
    return findIndex(t, k)
}
```

5. The next step is defining the `Decode` method. This will receive a pointer to a structure and then proceed to process lines from the scanner and populate the structure fields:

```
func (d *Decoder) Decode(v interface{}) error {
    val := reflect.ValueOf(v)
    t := val.Type()
    if t.Kind() != reflect.Ptr && t.Elem().Kind() != reflect.Struct
{
        return fmt.Errorf("%v not a struct pointer", t)
    }
    val = val.Elem()
    t = t.Elem()
    line := 0
    for d.scanner.Scan() {
        line++
        b := d.scanner.Bytes()
        if len(b) == 0 || b[0] == '#' {
            continue
        }
        parts := bytes.SplitN(b, []byte{':'}, 2)
        if len(parts) != 2 {
            return decodeError{line: line, err: errNoSep}
        }
        index, ok := findIndex(t, string(parts[0]))
        if !ok {
            continue
        }
        value := bytes.TrimSpace(parts[1])
        if err := d.decodeValue(val.Field(index), value); err !=
```

```
nil {
                return decodeError{line: line, err: err}
        }
    }
    return d.scanner.Err()
}
```

The most important part of the work will be done by the private `decodeValue` method. The first thing will be verifying that the `Unmarshaller` interface is satisfied, and, if it is, using it. Otherwise, the method is going to use reflection to decode the value received correctly. For each type, it will use a different `Set` method from `reflection.Value`, and it will return an error if it encounters an unknown type:

```
func (d *Decoder) decodeValue(v reflect.Value, value []byte) error {
    if v, ok := v.Addr().Interface().(Unmarshaller); ok {
        return v.UnmarshalProp(value)
    }
    switch valStr := string(value); v.Type().Kind() {
    case reflect.Int, reflect.Int8, reflect.Int16, reflect.Int32,
reflect.Int64:
        i, err := strconv.ParseInt(valStr, 10, 64)
        if err != nil {
            return err
        }
        v.SetInt(i)
    case reflect.Uint, reflect.Uint8, reflect.Uint16, reflect.Uint32,
reflect.Uint64:
        i, err := strconv.ParseUint(valStr, 10, 64)
        if err != nil {
            return err
        }
        v.SetUint(i)
    case reflect.Float32, reflect.Float64:
        i, err := strconv.ParseFloat(valStr, 64)
        if err != nil {
            return err
        }
        v.SetFloat(i)
    case reflect.String:
        v.SetString(valStr)
    case reflect.Bool:
        switch value := valStr; value {
        case "true":
            v.SetBool(true)
        case "false":
            v.SetBool(false)
        default:
```

```
            return fmt.Errorf("invalid bool: %s", value)
        }
    default:
        return fmt.Errorf("invalid type: %s", v.Type())
    }
    return nil
}
```

Using the package

To test whether the package behaves as expected, we can create a custom type that satisfies the Unmarshaller interface. The type implemented transforms the string into uppercase when decoding:

```
type UpperString string

func (u *UpperString) UnmarshalProp(b []byte) error {
        *u = UpperString(strings.ToUpper(string(b)))
        return nil
}
```

Now we can use the type as a structure field and we can verify that it gets transformed correctly in a decode operation:

```
func main() {
        r := strings.NewReader(
                "\n# comment, ignore\nkey1: 10.5\nkey2: some string" +
                        "\nkey3: 42\nkey4: false\nspecial: another
string\n")
        var v struct {
                Key1 float32
                Key2 string
                Key3 uint64
                Key4 bool
                Key5 UpperString `prop:"special"`
                key6 int
        }
        if err := prop.NewDecoder(r).Decode(&v); err != nil {
                log.Fatal(r)
        }
        log.Printf("%+v", v)
}
```

Summary

In this chapter, we reviewed the Go memory model for interfaces in detail, highlighting that an interface always contains a concrete type. We used this information to get a better look at type casting and understand what happens when an interface gets cast to another.

Then, we introduced the basic mechanics of reflection, starting with type and value, which are the two main types of the package. They represent, respectively, the type and value of a variable. Value allows you to read a variable content and also to write it if the variable is addressable. In order to be addressable, a variable needs to be accessed from its address, using a pointer, for instance.

We also saw how complex data types can be handled with reflection, seeing how to access structure field value. Data types of structure can be used to get metadata about fields, including name and tags, which are heavily used by encoding packages and other third-party libraries.

We saw how to create and operate with maps, including adding, setting, and deleting values. With slices, we saw how to edit their values and how to execute append operations. We also showed how to use a channel to send and receive data, and even how to use `select` statements in the same way we do with static-typed programming.

Finally, we listed where reflection is used in the standard library, and did a quick analysis of its computational cost. We concluded the chapter with some tips on how and when to use reflection in a library or in any application you are writing.

The next chapter is the last one of the book, and it explains how to leverage the existing C library in Go using CGO.

Questions

1. What's the memory representation of an interface in Go?
2. What happens when an interface type is cast to another one?
3. What are `Value`, `Type` , and `Kind` in reflection?
4. What does it mean if a value is addressable?
5. Why are structure field tags important in Go?
6. What's the general trade-off of reflection?
7. Could you describe a good approach to using reflection?

16
Using CGO

This chapter will introduce you to CGO, a Go runtime for the C language. It makes it possible to call C code from Go applications, and since C has a plethora of libraries available, this means that they can be leveraged in Go.

The following topics will be covered in this chapter:

- Using CGO from C and Go
- Understanding type differences

Technical requirements

This chapter requires Go to be installed and your favorite editor to be set up. For more information, refer to Chapter 3, *An Overview of Go*.

In addition, it requires the GCC compiler to be installed in your machine. This can easily be done on your Unix machine using the package manager. For Ubuntu, the command is as follows:

```
sudo apt install gcc
```

Introduction to CGO

CGO is the tool that makes it possible to run C code in a Go application. This feature has been around since Go reached version 1.0 in 2009 and allowed us to use existing C libraries when there were fewer packages available outside the standard library than today.

The C code is accessed through the C pseudo package, and it is accessed and called using the package name followed by the identifier, for instance, C.print.

The `import` declaration is preceded by a series of special comments, which specify what C source file the application should import:

```
package example

// #include <stdio.h>
import "C"
```

This statement can also be a multiline comment, which can contain more `include` directives, like the one from the example earlier, and even actual C code directly:

```
package example

/*
#include <stdio.h>
#include <stdlib.h>
#include <errno.h>

void someFunction(char* s) {
    printf("%s\n", s);
}
*/
import "C"
```

It is important to avoid blank lines between the C comment and the `import` statement, otherwise the libraries and the code will be imported by CGO in the application.

Calling C code from Go

To use existing C code, made by us or by someone else, we need to call C from Go. Let's perform a quick, complete example, where we are printing a string to the standard output using just C functionalities:

```
package main

/*
#include <stdio.h>
#include <stdlib.h>

void customPrint(char* s) {
    printf("%s\n", s);
}
*/
import "C"

import "unsafe"
```

```
func main() {
    s := C.CString(`Printing to stdout with CGO
        Using <stdio.h> and <stdlib.h>`)
            defer C.free(unsafe.Pointer(s))
            C.customPrint(s)
}
```

We are importing two C core libraries here, which are as follows:

- stdio.h : This contains the input and output methods. We are using printf.
- stdlib.h: This contains general functions, including memory management.

Looking at the preceding code, we notice that the variable that we are printing is not a normal Go string, but rather, it is obtained by the C.CString function that takes a string and returns a slice of char, because that's how strings are treated in C. The function is defined as follows:

```
func C.CString(string) *C.char
```

The second thing we can observe is that we are deferring a call to C.free, passing the s variable that we defined, but converted to a different type. This function call is essential since the language is not garbage collected and, in order to free the memory used, the application needs to specifically call the C free function. This function receives a generic pointer, which is represented by the unsafe.Pointer type in Go. According to the Go documentation, the following applies:

"A pointer value of any type can be converted into a Pointer."

This is exactly what we are doing, because the type of the string variable is the *C.char pointer.

Calling Go code from C

We just saw how to call C code from a Go application using the C package and the import statement. Now, we will see how to call Go code from C, which requires the use of another special statement called export. This is a comment that needs to be placed in the line above the function we want to export, followed by the name of that function:

```
//export theAnswer
func theAnswer() C.int {
    return 42
}
```

The Go function needs to be declared as external in the C code. This will allow the C code to use it:

```
extern int theAnswer();
```

We can test this functionality by creating a Go app that exports a function, which is used by a C function. This gets called inside the Go `main` function:

```
package main

// extern int goAdd(int, int);
//
// static int cAdd(int a, int b) {
//      return goAdd(a, b);
// }
import "C"
import "fmt"

//export goAdd
func goAdd(a, b C.int) C.int {
    return a + b
}

func main() {
    fmt.Println(C.cAdd(1, 3))
}
```

We can see in the preceding example that we have the `goAdd` function, which is exported to C with the `export` statement . The export name matches the name of the function, and there are no blank lines between the comment and the function.

We can notice that the types used in the signature of the exported function are not regular Go integers, but `C.int` variables. We will see how the C and Go systems differ in the next section.

The C and Go type systems

In order to pass data around between C and Go, we need to pass the correct types by executing the right conversion.

Strings and byte slices

The `string` type, which is a basic type in Go, does not exist in C. It has the `char` type, which represents a character, similar to Go's `rune` type, and strings are represented by an array of the `char` type, which is terminated with a `\0`.

The language makes it possible to declare character arrays directly as an array or as a string. The second declaration does not end the `0` value in order to end the string:

```
char lang[7] = {'G', 'o', 'l', 'a', 'n', 'g', '\0'};

char lang[] = "Golang";
```

We already saw how to convert a Go string to a C character array using the following function:

```
func C.CString(string) *C.char
```

This function will allocate the string in the heap so that it will be the application's responsibility to free such memory using the `C.free` function.

In order to convert a slice of bytes to a C character pointer named `*char`, we can use the following function:

```
func C.CBytes([]byte) unsafe.Pointer
```

As it happens, for `C.CString`, the application allocates the data in the heap and leaves the responsibilities of freeing it to the Go application.

The main difference between these two functions is that the first produces `char[]`, while the other creates `*char`. These two types are the equivalent of the Go `string` and `[]byte`, since the bytes of the first type cannot be changed, while the one from the second type can.

There are a series of functions that are used to convert the C types back to Go ones. As far as strings are concerned, there are two functions: `C.GoString` creates a string from the entire array, and `C.GoStringN` enables the creation of a string using an explicit length:

```
func C.GoString(*C.char) string

func C.GoStringN(*C.char, C.int) string
```

To transform the C `*char` back to Go `[]byte`, there is a single function:

```
func C.GoBytes(unsafe.Pointer, C.int) []byte
```

We can use the C.CBytes function to modify a slice of bytes using C and convert it back to a Go slice:

```
package main

/*
#include <stdio.h>
#include <stdlib.h>
#include <string.h>

char* reverseString(char* s) {
    int l = strlen(s);
    for (int i=0; i < l/2; i++) {
        char a = s[i];
        s[i] = s[l-1-i];
        s[l-1-i] = a;
    }
    return s;
}
*/
import "C"

import (
    "fmt"
    "unsafe"
)

func main() {
    b1 := []byte("A byte slice")
    c1 := C.CBytes(b1)
    fmt.Printf("Go ptr: %p\n", b1)
    fmt.Printf("C ptr: %p\n", c1)
    defer C.free(c1)
    c2 := unsafe.Pointer(C.reverseString((*C.char)(c1)))
    b2 := C.GoBytes(c2, C.int(len(b1)))
    fmt.Printf("Go ptr: %p\n", b2)
    fmt.Printf("%q -> %q", b1, b2)
}
```

Executing this application will show that when converting the byte slice, b1, to the C type as the c1 variable, it will change address. The C slice returned by the C function, c2, will have the same address as c1 because it is the same slice. When converted back to Go again and assigned to b2, it will have another address that is different from the initial Go byte slice, b1.

We can achieve the same result using the C string function. Let's use the same C code from the previous example and change the rest:

```
package main

/*
#include <stdio.h>
#include <stdlib.h>
#include <string.h>

char* reverseString(char* s) {
    int l = strlen(s);
    for (int i=0; i < l/2; i++) {
        char a = s[i];
        s[i] = s[l-1-i];
        s[l-1-i] = a;
    }
    return s;
}
*/
import "C"

import (
    "fmt"
    "unsafe"
)

func main() {
    s1 := "A byte slice"
    c1 := C.CString(s1)
    defer C.free(unsafe.Pointer(c1))
    c2 := C.reverseString(c1)
    s2 := C.GoString(c2)
    fmt.Printf("%q -> %q", s1, s2)
}
```

It is important to note that when transferring the Go string and bytes values to C, the values are copied. As a consequence, the C code is not capable of editing them directly, but will edit the copy, leaving the original Go value intact.

Integers

In C, the types of integers available have a number of similarities with Go, since there are signed and unsigned versions of each integer type in both languages, but they differ in terms of name and byte size. The C `sizeof` function makes it possible to check the size for each of these types.

Here is a list of integer types that are available in C:

Signed types

Type	Size	Range
char	1 byte	[-128, +127]
int	2 or 4 bytes	see short and long
short	2 bytes	[-32 768, +32 767]
long	4 bytes	[-2 147 483 648, +2 147 483 647]
long long	8 bytes	[-9 223 372 036 854 775 808, +9 223 372 036 854 775 807]

Unsigned types

Type	Size	Range
unsigned char	1 byte	[0, +255]
unsigned int	2 or 4 bytes	see unsigned short or unsigned long
unsigned short	2 bytes	[0, +65 535]
unsigned long	4 bytes	[0, +4 294 967 295]
unsigned long long	8 bytes	[0, +18 446 744 073 709 551 615]

The size of `int` in C depends on the architecture—it used to be 2 bytes with 16-bit processors, but with modern processors (32- and 64-bit), it's 4 bytes.

When we move from the realm of Go to the realm of C, and vice versa, we lose all variable overflow information. The compiler will not warn us when we try to fit an integer variable into another one that does not have an adequate size. We can see this with a brief example, as follows:

```
package main

import "C"

import "fmt"

func main() {
    a := int64(0x1122334455667788)

    // a fits in 64 bits
    fmt.Println(a)
    // short overflows, it's 16
    fmt.Println(C.short(a), int16(0x7788))
    // long also overflows, it's 32
    fmt.Println(C.long(a), int32(0x55667788))
    // longlong is okay, it's 64
    fmt.Println(C.longlong(a), int64(0x1122334455667788))
}
```

We can see that the value of `a` is a certain number, but the `short` and `long` variables do not have sufficient bytes, so they will have different values. The conversion shows that only the last bytes are taken from the variable when converting, and the other bytes are discarded.

Here is a useful list of C types and comparable Go types, as well as how to use them in Go code:

C type	Go Type	CGO type
char	int8	C.char
short	int16	C.short
long	int32, rune	C.long
long long	int64	C.longlong
int	int	C.int
unsigned char	uint8, byte	C.uchar
unsigned short	uint16	C.ushort
unsigned long	uint32	C.ulong
unsigned long long	uint64	C.ulonglong
unsigned int	uint	C.uint

You can use this table as a reference when performing conversions and avoid errors derived from using the wrong type, since there are no overflow warnings when using CGO.

Float types

In C, the `float` types are very similar to the Go ones:

- C offers `float`, which is 32 bit, and `double`, which is 64 bit.
- Go has `float32` and `float64`.

This can cause rounding errors when converting from a 64-bit value to a 32-bit one, as shown in the following code:

```
package main

import "C"

import (
    "fmt"
    "math"
)

func main() {
    a := float64(math.Pi)

    fmt.Println(a)
    fmt.Println(C.float(a))
    fmt.Println(C.double(a))
    fmt.Println(C.double(C.float(a)) - C.double(a))
}
```

The preceding example shows how the `math.Pi` value goes from `3.141592653589793` to `3.1415927`, causing an error of about $1/10^7$.

Unsafe conversions

We will now see how it is possible to edit a Go variable from C using the `unsafe` package.

Editing a byte slice directly

It also possible to edit a Go byte slice directly using a dirty trick. From Go's perspective, a slice is a trio of values:

- A pointer to the first element
- The size of slice
- The capacity of the slice

In C, a byte slice is just a series of bytes, and a string is a character slice that ends with a \0.

If we use the unsafe package to pass the pointer to the first element of the slice, we will be able to edit the existing byte slice directly without executing copies and transformations. We can see how to execute this conversion in the following application:

```
package main

/*
#include <stdio.h>
#include <stdlib.h>
#include <string.h>

void reverseString(char* s) {
    int l = strlen(s);
    for (int i=0; i < l/2; i++) {
        char a = s[i];
        s[i] = s[l-1-i];
        s[l-1-i] = a;
    }
}
*/
import "C"

import (
  "fmt"
  "unsafe"
)

func main() {
    b1 := []byte("A byte slice")
    fmt.Printf("Slice: %s\n", b1)
    C.reverseString((*C.char)(unsafe.Pointer(&b1[0])))
    fmt.Printf("Slice: %s\n", b1)
}
```

The conversion is executed using the expression `(*C.char)(unsafe.Pointer(&b1[0]))`, which does the following:

- Takes the pointer to element zero for the slice
- Converts it into an unsafe pointer
- Converts the `byte` pointer into a `C.char` pointer, which shares the memory representation

Numbers

Using the `unsafe` package, we can also convert a numeric variable pointer to its C counterpart. This allows us to edit it directly in the C code:

```
package main

/*
void half(double* f) {
    *f = *f/2;
}
*/
import "C"

import (
    "fmt"
    "math"
    "unsafe"
)

func main() {
    a := float64(math.Pi)
    fmt.Println(a)
    C.half((*C.double)(unsafe.Pointer(&a)))
    fmt.Println(a)
}
```

The preceding example does exactly that; it halves the value of `a` in a C function, without copying and assigning the new value in Go.

Working with slices

Go slices and C slices differ in one fundamental aspect—the Go version embeds both length and capacity, while in C, all we have is a pointer to the first element. This means that in C, length and capacity must be stored somewhere else, such as in another variable.

Let's take the following Go function, which calculates the mean of a series of `float64` numbers:

```
func mean(l []float64) (m float64) {
    for _, a := range l {
        m += a
    }
    return m / float64(len(l))
}
```

If we want to have a similar function in C, we need to pass a pointer together with its length. This will avoid errors such as segmentation fault, which happens when an application tries to gain access to memory that has not been assigned to it. If the memory is still assigned to the application, the result is that it provides access to a memory area with an unknown value, causing unpredictable results:

```
double mean(int len, double *a) {
    if (a == NULL || len == 0) {
        return 0;
    }
    double m = 0;
    for (int i = 0; i < len; i++) {
        m+=a[i];
    }
    return m / len;
}
```

We can try this function using a Go wrapper that takes a slice and also passes the length to the C function:

```
func mean(a []float64) float64 {
    if len(a) == 0 {
        return 0
    }
    return float64(C.mean(C.int(len(a)), (*C.double)(&a[0])))
}
```

To verify what happens, we can also create a similar function that passes a length that is not correct:

```
func mean2(a []float64) float64 {
    if len(a) == 0 {
        return 0
    }
    return float64(C.mean(C.int(len(a)*2), (*C.double)(&a[0])))
}
```

When using this function, we will see that the application should not raise any segmentation fault error, but the result obtained will be different. This is because the second one will add a series of extra values to the mean calculation, as shown here:

```
var a = make([]float64, 10)

func init() {
    for i := range a {
        a[i] = float64(i + 1)
    }
}

func main() {
    cases := [][]float64{a, a[1:4], a[:0], nil}
    for _, slice := range cases {
        fmt.Println(slice, mean(slice))
    }
    for _, slice := range cases {
        fmt.Println(slice, mean2(slice))
    }
}
```

Working with structs

After seeing how slices work, we will know how to handle complex data in C and Go using structures. For this, let's see the following sections.

Structures in Go

Go structures use a technique called alignment, which consists of adding one or more bytes to data structures to make it fit into memory addresses better. Consider the following data structure:

```
struct {
    a string
    b bool
    c []byte
}
```

With 64-bit architecture calling `unsafe.Sizeof` on this structure, this will give us an unexpected result. What we are expecting is the following:

- 16 bytes from the string; 8 for the pointer to the first element, and 8 for the length
- 1 byte for the Boolean
- 24 for the slice; 8 for the address, 8 for the length, and 8 for capacity

The total should be 41, but the function returns 48. This happens because the compiler inserts extra bytes after the Boolean, to reach 8 bytes (64 bits) and optimize the operation for the CPU. The structure could be represented in memory as follows:

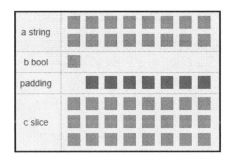

We can see that the Boolean variable takes 1 bit, and that there are 7 extra bits added by the compiler. This is very helpful because it avoids the other variables that are store, half in one memory slot and half in another. This would require two reads and two writes per operation, with a significant drop in performance.

If two or more fields are small enough to fit in one slot of 64 bits, they will be stored sequentially. We can see this with the following example:

```
struct {
    a, b bool
    c rune
    d byte
    e string
}
```

This structure translates into the following memory representation on a 64-bit architecture:

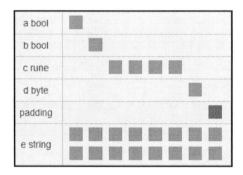

We can clearly see that both the Boolean variables, the `rune` and the `byte`, are in the same memory address, and a byte of padding is added to align the last field.

Manual padding

Go makes it possible to specify padding in structs manually using the blank identifier for fields. Take the following data structure:

```
struct{
    a int32
    b int32
}
```

This will have the following representation:

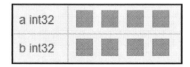

We can use the blank identifier to specify padding manually and to optimize the data structure for a 64-bit architecture:

```
struct{
    a int32
    _ int32
    b int32
    _ int32
}
```

This will allow the application to store each `int32` in its own memory location because the blank fields will be acting as padding:

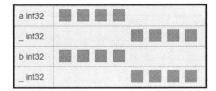

Structures in C

Structures in C share the same concept of alignment that Go uses, but they are always aligned using 4-byte padding. Unlike Go, it is possible to avoid padding entirely, which helps to save space by reducing memory usage. Let's find out more about this in the following sections.

Unpacked structures

Each struct we define will be unpacked unless otherwise specified. We can define a structure in C as follows:

```
typedef struct{
  unsigned char a;
  char b;
  int c;
  unsigned int d;
  char e[10];
} myStruct;
```

We can use it and populate it with values directly from our Go code without any issues arising:

```
func main() {
    v := C.myStruct{
        a: C.uchar('A'),
        b: C.char('Z'),
        c: C.int(100),
        d: C.uint(10),
        e: [10]C.char{'h', 'e', 'l', 'l', 'o'},
    }
    log.Printf("%#v", v)
}
```

This small test will give us the following output:

```
main._Ctype_struct___0{
    a:0x41,
    b:90,
    c:100,
    d:0xa,
    e:[10]main._Ctype_char{104, 101, 108, 108, 111, 0, 0, 0, 0, 0},
     _:[2]uint8{0x0, 0x0},
}
```

This tells us that there is an extra blank field that is used for padding because the last field is 10 bytes, which is 2 bytes short of being a multiple of 4 (that is, 12 bytes).

Packed structures

We can define a packed struct in C using the `pragma pack` directive. We can pack the structure from before as follows:

```
#pragma pack(1)
typedef struct{
  unsigned char a;
  char b;
  int c;
  unsigned int d;
  char e[10];
} myStruct;
```

If we try to use the C structure in our Go code, we will obtain a compile error, if using fields c and d:

```
pack := C.myStruct{
    a: C.uchar('A'),
    b: C.char('Z'),
    c: C.int(100),
    d: C.uint(10),
    e: [10]C.char{},
}
```

If we try to print the structure as we did for the unpacked version, we will see the reason why:

```
main._Ctype_struct___0{
    a:0x41,
    b:90,
    _:[8]uint8{0x0, 0x0, 0x0, 0x0, 0x0, 0x0, 0x0, 0x0},
    e:[10]main._Ctype_char{104, 101, 108, 108, 111, 0, 0, 0, 0, 0},
}
```

We can see from the output that the c and d fields, both of 4 bytes, are replaced by one empty field of 8 bytes that Go cannot access. Consequently, we cannot populate the structure from Go, but we can access this field in the C part of the application:

```
myStruct makeStruct(){
  myStruct p;
  p.a = 'A';
  p.b = 'Z';
  p.c = 100;
  p.d = 10;
  p.e[0] = 'h';
  p.e[1] = 'e';
  p.e[2] = 'l';
  p.e[3] = 'l';
  p.e[4] = 'o';
  p.e[5] = '\0';
  p.e[6] = '\0';
  p.e[7] = '\0';
  p.e[8] = '\0';
  p.e[9] = '\0';
  return p;
}
```

This will allow us to return a structure with the correct values. We can print it and see that the _ field contains both the values of c and d:

```
main._Ctype_struct___0{
    a:0x41,
    b:90,
    _:[8]uint8{0x64, 0x0, 0x0, 0x0, 0xa, 0x0, 0x0, 0x0},
    e:[10]main._Ctype_char{104, 101, 108, 108, 111, 0, 0, 0, 0, 0}
}
```

Now that we have the data, we need to create a Go structure that can host it:

```
type myStruct struct {
    a uint8
    b int8
    c int32
    d uint32
    e [10]uint8
}
```

Now, we need to read the raw bytes from the C structure and unpack it manually:

```
func unpack(i *C.myStruct) (m myStruct) {
    b := bytes.NewBuffer(C.GoBytes(unsafe.Pointer(i), C.sizeof_myStruct))
    for _, v := range []interface{}{&m.a, &m.b, &m.c, &m.d, &m.e} {
        binary.Read(b, binary.LittleEndian, v)
    }
    return
}
```

We can use the C.GoBytes function, which works for any pointer (not just bytes) and specifies the size of the struct we defined, which is stored in the constant, C.sizeof_myStruct. We can then proceed to read each field in order using the binary.Read function with **Little-Endian (LE)** encoding.

We can see that the resulting structure contains all the data in the correct fields:

```
main.myStruct{
    a:0x41,
    b:90,
    c:100,
    d:0xa,
    e:[10]uint8{0x68, 0x65, 0x6c, 0x6c, 0x6f, 0x0, 0x0, 0x0, 0x0, 0x0},
}
```

CGO recommendations

We have now seen how to use CGO with integers, floating pointers, slices, and structures. This is a very powerful tool that facilitates the use of a lot of the existing C code in our Go applications. As we did for reflection in the previous chapter, we are now going to talk about the less obvious downsides of CGO.

Compilation and speed

One of Go's trademarks is blazingly fast compilation times. When using CGO, the amount of work involved in the compilation is much higher, and involves more than just passing all the `.go` files to the Go compiler. The compilation process goes more or less as follows:

- CGO needs to create C to Go and Go to C stubs.
- The `make` command needs to be called to compile all the C source files.
- All the files are combined in a `.o` file.
- The system's linker needs to verify that all the references between Go and C are valid.

If this process goes smoothly, you can launch your application, but if you encounter any issues, you will need to check the errors between C and Go, which is not as easy as debugging a pure Go application.

Another drawback is that not every operating system comes with a `make` command out of the box. The C part may require some extra flags to compile correctly, and that cannot be handled by `go install` or `go build`. You will need to create a compilation script for your application, such as a `makefile` script.

Performance

While discussing how to let C and Go talk to one another, we saw that for each type, there is a conversion operation that needs to be executed. This can be straightforward, as it happens, for numbers, but a little bit more complex for strings, bytes, or slices, and even more when we talk about structures. These operations are not cost-free, both in terms of memory usage and performance. This will not be a problem for many applications, but if you're trying to achieve high performance, it could be your bottleneck.

The C code is not aware of what's happening in its Go counterpart. When it needs to be invoked, Go needs to pass the information about its stack to C in a format that is suitable for it. When the C code finishes its execution, the information about the stack state and the variables used needs to be transferred back from C to Go.

Dependency from C

When using CGO, you face the same problem that other languages face when creating bindings or wrappers to C code. You are completely dependent on it.

The Go application has to handle the way C uses memory and other resources, while the C application is not aware of what Go is doing and is not using any kind of concurrency, neither goroutines nor threads.

In addition to that, C code is hard to debug, maintain, and replace if you're not a C developer. This way, sometimes it is better to write a library from the ground up, instead of relying on the existing C implementation.

A very good example of this is `go-git` (`https://github.com/src-d/go-git`), which implements the Git protocol functionality in pure Go by imitating the existing C library, `libgit2`.

Summary

In this chapter, we saw a very powerful tool in the Go arsenal: CGO. This allows a Go application to run C code, which, in turn, can invoke Go functions. We saw that it requires a special `import` statement, `import "C"`, which is a pseudo package that contains all the C code available to Go. To export Go code and make it available to C, there is a special comment, `//export`, which makes the Go function available in the C namespace.

We saw that the C and Go type systems are very similar for some things, but very different for others. We saw that strings and byte arrays can be converted to C types, and vice versa. Integers in C and Go are also pretty similar, the main difference being the `int` type. In C, this is 4 bytes, while in Go, it is 4 or 8 bytes, depending on the architecture. Floats are also pretty similar, with a 4- and 8-bit version in both C and Go, which just have different names.

It is also possible to edit a numeric Go variable or a byte slice directly, without creating a copy. This is possible using the `unsafe.Pointer` function to force a casting that wouldn't otherwise be allowed. Slices in C are just pointers to the first element, and the length of the slice needs to be stored in another variable. That's why we created Go functions that take the slice and pass both parameters to their C counterparts.

Before talking about data structures, we must mention what alignment is, how Go achieves it, and how C alignment differs from that of Go. Data structures in CGO use alignment, and are pretty straightforward to use. If they are not packed, we can pass them around and extract values very easily. If the structure is packed instead, we cannot access some of its fields and we need a workaround to manually execute a conversion to Go.

The final topic focused on the downsides of CGO, from its slower building time to its drop in performance because of the conversion that is required, and how the application is going to end up being more difficult to maintain because of the C code.

I hope you enjoyed this Go journey so far, and that it will help you to write modern, concurrent, and efficient applications.

Questions

1. What is CGO?
2. How can you call C code from Go?
3. How can you use Go code in C?
4. What difference in data types is there between Go and C?
5. How can you edit Go values inside C code?
6. What is the main problem with packed data structures?
7. What are the main downfalls of CGO?

Assessments

Chapter 1

1. **What is the difference between application and system programming?**
   Application programming focuses on solving a problem for the final user, while system programming is about creating software used by other software.

2. **What is an API? Why are APIs so important?**
   An API is an interface that the software exposes to control the access to the resources it controls. It is a description of how other applications should communicate with the software.

3. **Could you explain how protection rings work?**
   Protection rings are a system used to prevent failures and increase security. They arrange security in hierarchical levels with growing limitations and allow a mediate access to the features of the more powerful levels by using specific gateways.

4. **Can you provide some examples of what cannot be undertaken in user space?**
   An application in user space cannot change its current space to kernel, cannot access the hard drive ignoring the filesystem, and cannot change the page tables.

5. **What's a system call?**
   System calls are the API provided by the operating system to access the machine's resources.

6. **Which calls are used in Unix to manage a process?**
   The calls that Unix uses to manage a process are as follows: `fork`, `exit`, and `wait`.

7. **Why is POSIX useful?**
   The various POSIX standards define process controls, signals, segmentation, illegal instructions, file and directory operations, pipes, I/O control and the C library, shells and utilities, and real-time and multithreaded extensions. It is extremely useful for a developer when building applications, because it aids in building an application that works with different operating systems sharing this standard.

8. **Is Windows POSIX-compliant?**
   Windows is not POSIX-compliant, but an attempt is being made to offer a POSIX framework, such as the Windows Linux subsystem.

Chapter 2

1. **Which filesystem is used by modern operating systems?**
   Modern operating systems use different filesystems: Windows and macOS use their respective proprietary formats, NTFS and APFS, while Linux systems mainly use EXT4.

2. **What is an inode? What is inode 0 in Unix?**
   An inode is a filesystem data structure representing a file. It stores information about a file, excluding the name and data.
   The inode 0 is reserved for the / folder.

3. **What's the difference between PID and PPID?**
   PID is the unique identifier for an existing process, while PPID is the identifier of the parent process. When an existing process creates another, the new process has a PPID equal to the existing process's PID.

4. **How do you terminate a process running in the background?**
   While a SIGINT signal can be sent to a foreground process by pressing *Ctrl + C*, for a background process, the signal needs to be sent with the kill command, in this case, kill -2 PID.

5. **What is the difference between a user and a group?**
   A user identifies an account that can own files and processes, while a group is a mechanism to share permissions on a file.

6. **What's the scope of the Unix permission model?**
   The Unix permission model enables the restriction of access to a file with three different levels of power: owner, group, and all other users.

7. **Can you explain the difference between signals and exit codes?**
   Signals and exit codes are both methods of communication between processes, but whereas a signal is from any process to another, exit codes are used to communicate from a child to its parent.

8. **What's a swap file?**
   A swap file is an extension of the physical memory that is used to store pages that are not required in order to free up the main memory.

Chapter 3

1. **What's the difference between an exported and an unexported symbol?**
   Exported symbols can be used by other packages, whereas unexported symbols cannot. The first group has an identifier that starts with a capital letter, while the second group does not.

2. **Why are custom types important?**
   Custom types allow methods to be defined and interfaces to be used effectively, or the data structure of another type to be inherited, but getting rid of its methods.

3. **What is the main limit of a short declaration?**
   Short declarations do not allow the variable type inferred by the value to be defined. A type casting of the value enables this limitation to be overcome.

4. **What is scope and how does it affect variable shadowing?**
   The scope of a variable represents its lifetime and visibility that can be package, function, or blocked. Shadowing is when the same identifier gets used in an inner scope, preventing access to the symbol that shares that identifier by an outer scope.

5. **How can you access a method?**
   Methods are special types of functions that have a namespace linked to the type they belong to. They can be accessed as a property of an instance of their type, or as a property of the type itself, passing the instance as a first argument.

6. **Explain the difference between a series of** `if`/`else` **statements and a** `switch` **statement.**
   A series of `if` and `else` statements allow the execution of a short declaration for each `if` statement and will execute only one of the cases, skipping the following declarations. A `switch` statement allows only one declaration, and can modify the flow using `continue` and `break` statements.

7. **In a typical use case, who is generally responsible for closing a channel?**
   Channels should always be closed by the sender, because that party is responsible for communicating that there is no more information to send. Also, sending to a closed channel throws a panic, while receiving from it is a non-blocking operation.

8. **What is escape analysis?**
   Escape analysis is an optimization process performed by the Go compiler that attempts to reduce the variables allocated in the heap by verifying whether they outlive the function where they are defined.

Chapter 4

1. **What's the difference between absolute and relative paths?**
   An absolute path starts with the / (root) path, while a relative path does not. To obtain an absolute path from a relative one, it must be joined to the current working directory.
2. **How do you obtain or change the current working directory?**
   To find out the current working directory, the os package offers the Getwd function, which returns the current working directory. To change the current working directory, the Chdir function must be used. It accepts both relative and absolute paths.
3. **What are the advantages and downfalls of using** ioutil.ReadAll**?**
   The ioutil.ReadAll function places the entire file contents in a byte slice, so the size of file influences the amount of memory allocated, and then released. Since there is no recycling of the memory allocated this way, these slices get garbage-collected when they are no longer used.
4. **Why are buffers important for reading operations?**
   Byte buffers limit the amount of memory allocated by the reading operations, but they also require a certain number of read operations, each one with a little overhead that impacts speed and performance.
5. **When should you use** ioutil.WriteFile**?**
   The ioutil.WriteFile function can be used if the size of the content is not too big, because the entire content needs to be in memory. It is also preferable to use it in short-lived applications and avoid it for recurrent writing operations.
6. **Which operations are available when using a buffered reader that allows peeking?**
   The peeking operation allows the content of the next bytes to be checked without advancing the cursor of the current reader, and this enables us to have contextual operations, such as read word, read line, or any custom token-based operation.
7. **When is it better to read content using a byte buffer?**
   Using a reading buffer is a way of lowering the memory usage of your application. It can be used when there's no need to have all the content at once.
8. **How can buffers be used for writing? What's the advantage of using them?**
   In writing operations, the application already handles the bytes that are about to be written, so an underlying buffer is used to optimize the number of system calls, only when the buffer is full, so as to avoid the addition of system call overheads when the data passed to the writer is not enough.

Chapter 5

1. **What's a stream?**
   A stream is an abstraction that represents a generic flow of incoming or outgoing data.

2. **What interfaces abstract the incoming streams?**
   The `io.Reader` interface is an abstraction for incoming streams.

3. **Which interface represents the outgoing streams?**
   The `io.Writer` interface is an abstraction for outgoing streams.

4. **When should a byte reader be used? When should a string reader be used instead?**
   A byte reader should be used when the raw data is a slice of bytes, while a string reader should be used with strings. Converting from one type of data to another causes a copy and is inconvenient.

5. **What's the difference between a string builder and a byte buffer?**
   A byte buffer can be reused and overwritten. A string builder is used to create a string without a copy, so it uses a byte slice and converts it to a string without copying, using the `unsafe` package.

6. **Why should reader and writer implementations accept an interface as input?**
   Accepting an interface as an input means to be open to different types with the same behavior. This enables existing readers and writers, such as buffers and files, to be used.

7. **How does a pipe differ from** `TeeReader`**?**
   A pipe connects a writer to a reader. Whatever gets written in the writer gets read by the reader. A `TeeReader` does the opposite, connecting a reader to a writer, so what gets read is also written somewhere else.

Chapter 6

1. **What is a Terminal, and what is a pseudo-terminal?**
   A terminal is an application that behaves like a teletype, by displaying a 2 x 2 matrix of characters. Pseudo terminals are applications that run under a terminal and emulate its behavior by being interactive.

2. **What should a pseudo terminal be able to do?**
   A pseudo terminal application should be able to receive input from a user, execute an action according to the instruction received, and display the result back to the user.

3. **What Go tools did we use in order to emulate a terminal?**
   To manage user input, we used a buffered scanner in standard input, which will read user input line by line. Each command has been implemented using the same interface. To understand the command invoked, we used a comparison between the first argument and the commands available. A writer is passed to command to print their output.

4. **How can my application get instructions from the standard input?**
   The application can use the standard input combined with a scanner that will return a new token each time it encounters a new line.

5. **What is the advantage of using interfaces for commands?**
   Using interfaces for commands allows us and the user of our package to expand the behavior by implementing their own version of the interface.

6. **What is the Levenshtein distance? Why can it be useful in pseudo-terminals?**
   The Levenshtein distance is the number of changes required to transform a string into another. It can useful for suggesting other commands to the user when they specify a non-existing one.

Chapter 7

1. **What applications are available for the current process inside a Go application?**
   The applications available for a process are PID (process ID), PPID (parent PID), UID and GID (user and group ID), and the working directory.

2. **How do you create a child process?**
   The `exec.Cmd` data structure can be used to define a child process. The process gets created when one of the `Run`, `Start`, `Output`, and `CombinedOutput` methods gets called.

3. **How do you ensure that a child process survives its parent?**
   By default in Unix systems, a child survives if the parent terminates. Additionally, you can change the process group and session ID of the child in order to ensure that it survives the parent.

4. **Can you access child properties? How can they be used?**
   One of the biggest advantages is to access the child PID to persist it somewhere, such as on the disk. This will allow another instance of the application, or any other application, to know which is the identifier of the child and verify whether it's still running.

5. **What's a daemon in Linux and how are they handled?**
A daemon in Linux is a process that is running in the background. In order to create a daemon, a process can create a fork of itself and terminate, set the init process to be the parent of the fork, set the current working directory to root for the fork, setting the input of the child to null, and use log files for output and error.

Chapter 8

1. **What's an exit code? Who makes use of it?**
The exit code is an integer value passed from a process to is parent to signal when the process ends. It represents the outcome of the process, and it is 0 if there have been no errors. The parent process can use this value to decide what to do next, such as running the process again if there is an error.

2. **What happens when an application panics? What exit code is returned?**
If panic is not recovered, the application will execute all the deferred functions and will exit with a status of 2.

3. **What's the default behavior of a Go application when receiving all signals?**
The default behavior of a Go application with signals is an early exit.

4. **How do you intercept signals and decide how the application must behave?**
The signals received can be intercepted using the signal.Notify method on a channel, specifying the type of signals that you want to handle. The values received by the channel can be compared to signal values, and the application can behave accordingly.

5. **Can you send signals to other processes? If so, how?**
It is possible to send signals to another process inside a Go application. In order to do so, the application needs to acquire an instance of the os.Process structure using a lookup function, and then it can use the Signal method of the structure to send a signal.

6. **What are pipes and why are they important?**
Pipes are two streams, one of output and the other of input, connected together. What's written in the output is available to the input, and this facilitates the connection of one process output to another process input.

Chapter 9

1. **What's the advantage of using communication models?**
   Communication models allow you to abstract the type of data handled with your model, making the communications between different endpoints easy.

2. **What's the difference between a TCP and a UDP connection?**
   TCP is connection oriented—this makes it reliable because it verifies that the destination receives data correctly before sending new data. A UDP connection sends data continuously, without acknowledging that the destination received the package. This can cause package loss, but it makes the connection faster and does not accumulate latency.

3. **Who closes the request body when sending requests?**
   Closing the request when making an HTTP call is the responsibility of the application.

4. **Who closes the body when receiving requests in the server?**
   The request body is closed automatically when the connection is closed, but the server can close it even earlier if it so desires.

Chapter 10

1. **What's the trade-off between text and binary encodings?**
   Text-based encodings are easier to read for a human, as well as easier to debug and write, but they take more space because of it. Binary encodings are difficult to write, read, and debug for a human, but smaller in size.

2. **How does Go behave with a data structure by default when encoding?**
   The default behavior of Go is to use reflection in order to read the fields and their values.

3. **How can this behavior be changed?**
   This behavior can be changed by implementing the marshaller interface of the encoding you are using, such as `json.Marshaller` for JSON.

4. **How does a structure field get encoded in an XML attribute?**
   The struct field needs to specify the `,attr` value in its tag.

5. **What operation is required to decode a `gob` interface value?**
   The data types that implement the interface need to be registered in the `gob` package using the `gob.Register` function.

6. **What is the protocol buffer encoding?**
The protocol buffer is an encoding protocol made by Google that uses a definition file for data structures and services. The file is used to generate data models, clients, and server stubs, leaving only the implementation of the server to the developer.

Chapter 11

1. **What is a thread and who is responsible for it?**
A thread is a part of a process that can be assigned by a specific core or CPU. It carries information about the state of the application, like a process does, and is managed by the operating system scheduler.

2. **How do goroutines differ from threads?**
Goroutines are tiny in size compared to threads, with a 1 to 100 ratio, and they are not managed by the operating system. The Go runtime takes care of the scheduling of goroutines.

3. **When are arguments evaluated when launching a goroutine?**
All the arguments passed to the function that starts the goroutine are evaluated when the goroutine is created. This means that if the value of the argument changes before the goroutine actually gets picked up by the scheduler and starts, the change is not going to be reflected in the goroutine.

4. **How do buffered and non-buffered channels differ?**
A non-buffered channel is created by the `make` function if no capacity is specified, or if it's 0. Each send operation to such a channel will block the current goroutine, until a receive operation is performed by another goroutine. A buffered channel can support a number of non-blocking send operations equal to its capacity. This means that if a channel has a capacity of n, the first n-1 send operations that are not matched by any receive operation will not be blocking.

5. **Why are one-way channels useful?**
They allow just a subset of operations, making clear to the user what the scope of the channel is. A receive-only channel does not allow data to be sent, or to close it, and that makes perfect sense because it is not the responsibility of the receiver. A send-only channel does not allow the receipt of data, but allows it to be sent and to close the channel, with an implicit statement that it is up to the sender to close the channel to signal that there is no more data.

6. **What happens when operations are executed on** `nil` **or closed channels?**
Sending to, or receiving from, a `nil` channel blocks forever, and closing it creates panics. Receiving from a closed channel returns a zero value immediately, and `false`, while sending to a closed channel raises a panic and the same thing happens if we try to close it again.

7. **What are timers and tickers used for?**
Timers and tickers both create a receive-only channel. Timers can be used in a loop with a `select` statement, instead of using `default`, in order to reduce the frequency of the selection and lower the CPU usage of the application when it is idle. Tickers are very useful for executing an operation every fixed period of time, while one practical usage is a rate limiter, which limits the number of executions over a set period of time in a certain segment of an application.

Chapter 12

1. **What's a race condition?**
A race condition is a situation where an application tries to execute two operations on the same resource at the same time, and the nature of the resource only allows one operation at time.

2. **What happens when you try to execute read and write operations concurrently with a map?**
When reading and writing operations on a map happen simultaneously, this causes a runtime error: `concurrent map writes`.

3. **What's the difference between** `Mutex` **and** `RWMutex`**?**
A regular mutex allows a resource to be locked and unlocked, and each operation has the same priority. A read/write mutex has two types of locks, one for each operation (read/write). The read lock allows more than one operation at time, while it is exclusive. Write locks could be subject to a delay if there are many continuous read operations on the resource. This is known as write starvation.

4. **Why are wait groups useful?**
Wait groups are the perfect tool to synchronize with the execution of different goroutines. This enables a clean and elegant solution to the classic setting, where there are several concurrent operations, and a main goroutine has to wait for them to end before moving on.

5. **What's the main use of** `sync.Once`**?**
`sync.Once` can be used to execute a concurrent operation on one occasion. It can be used to close a channel once and avoid panics, for instance. Another use case is the lazy initialization of a variable to implement a thread-safe version of the singleton design pattern.

6. **How can you use a pool?**
   A pool allows short-lived items to be reused. A good use case for pools is byte slices and byte buffers, because the pool will prevent this resource from being recycled by the garbage collector, while preventing the allocation of new pools.

7. **What's the advantage of using atomic operations?**
   Using a mutex for numeric variables has a lot of overhead. Atomic operations allow such overheads to be reduced and thread-safe operations to be executed on numeric variables. Its main use is for integer numbers, but, with some transformation, we can do the same for other types, such as Booleans and floats.

Chapter 13

1. **What is context in Go?**
   Context is a package that contains a generic interface and some auxiliary functions to return context instances. It is used to synchronize operations between various parts of the application and to carry values.

2. **What's the difference between cancellation, deadline, and timeout?**
   There are three different types of expiration for a context—cancellation is an explicit call to a cancellation function by the application, deadline is when the context goes over a specified time, and timeout is when the context survives a specific duration.

3. **What are the best practices when passing values with a context?**
   Values passed around with context should be relative to the current scope or request. They should not be used as a way of passing around optional function parameters or variables that are essential to the application. It is also a good idea to use custom private types as keys, because built-in values can be overridden by other packages. Pointers to values are also a solution to such a problem.

4. **Which standard packages already use context?**
   There are different packages that use context. The most notable are `net/http`, which uses context for requests and for server shutdown; `net`, which uses context for functions such as `Dial` or `Listen`; and `database/sql`, which uses the context as a way to cancel operations such as queries.

Chapter 14

1. **What is a generator? What are its responsibilities?**
   A generator is a tool that returns a series of values—it returns the next value in the series each time it is called. It's responsible for generating values in the sequence on demand. In Go, this can be done by using a channel to receive the values that are sent through by a goroutine that creates them.

2. **How would you describe a pipeline?**
   A pipeline is a type of application flow that splits the execution into different stages. These stages communicate with one another by using a certain means of communication, such as networks, or runtime internals, such as channels.

3. **What type of stage gets a channel and returns one?**
   An intermediate stage will receive from a receive-only channel and return another receive-only channel.

4. **What is the difference between fan-in and fan-out?**
   Fan-in is also known as demultiplexing, and entails gathering messages from different sources into one. Fan-out, or multiplexing, is the opposite—it entails splitting a single source of a message to more receivers.

Chapter 15

1. **What's the memory representation of an interface in Go?**
   An interface in Go is represented by two values—the first one is the interface concrete type, while the second is the value for such a type.

2. **What happens when an interface type is casted to another one?**
   Since interface values need to be a concrete value, and cannot be another interface, a new interface is created with a different type and the same concrete value.

3. **What are** Value, Type, **and** Kind **in reflection?**
   A Value, as the name suggests, represents the content of a variable; a Type represents the Go type of a variable; and Kind is the memory representation of a Type and refers only to built-in types.

4. **What does it mean that a value is addressable?**
   An addressable value is a value that can be edited because it has been obtained by a pointer.

5. **Why are structure field tags important in Go?**
   Structure field tags are an easy way to add extra information about a structure field that is easy to read, using the reflection Type interface.

6. **What's the general trade-off of reflection?**
   Reflection allows your code to deal with unknown types of data and make your package or application generic, but it comes with an overhead that has a performance cost. It also makes code more obscure and less maintainable.

7. **Could you describe a good approach when using reflection?**
   The best approach to reflection is the one that we find in many different parts of the standard library; for instance, in the `encoding` packages. They use reflection as a last resort, and they do so by providing interfaces for encoding and decoding operations. If these interfaces are satisfied by a type, the package will use the respective methods instead of relying on reflection.

Chapter 16

1. **What is CGO?**
   CGO is a powerful Go tool that handles communication between C code and Go code. This allows C code to be used in a Go application and to leverage the huge amount of existing C libraries.

2. **How can you call C code from Go?**
   Go offers a pseudo package called `C` that exposed C types, such as `C.int`, and some functions that will convert Go strings and bytes into `C` character arrays, and vice versa. The comment that comes before the import `C` package will be interpreted as C code, and all the functions defined in it (be it directly, or by importing files), will be available in Go as functions of the `C` package.

3. **How can you use Go code in C?**
   If a Go function is preceded by a special comment, `//export`, this function will be available to the C code. It will also have to be defined as an external function in C.

4. **What are the differences in data types between Go and C?**
   Even if they have different data types, C and Go share most of their built-in numeric types. Strings in Go are a built-in immutable type, but in C, they are just a character array terminated by a `\0` value.

5. **How can you edit Go values inside C code?**
   Using the `unsafe` package, you can convert data types with the same memory representation in both C and Go. You need to convert the pointer to a value in its C counterpart, and this will allow you to edit the pointer content from the `C` part of the application.

6. **What is the main problem associated with packed data structures?**
   Packed data structures save space in memory, but their fields can be unaligned, meaning that they are split between multiple memory zones. This means that read and write operations take twice as long. There is also another inconvenience—some of the packed fields are not directly accessible from Go.

7. **What are the main downfalls of CGO?**
   Even if it is a very powerful tool, CGO has many downsides—the performance cost of passing from C to Go, and vice versa; the fact that the compiling time increases because the C compiler gets involved in the process; and that your Go code is reliant on your C code to work, which could be harder to maintain and debug.

Other Books You May Enjoy

If you enjoyed this book, you may be interested in these other books by Packt:

Hands-On Full Stack Development with Go

Mina Andrawos

ISBN: 978-1-78913-075-1

- Understand Go programming by building a real-world application
- Learn the React framework to develop a frontend for your application
- Understand isomorphic web development utilizing the GopherJS framework
- Explore methods to write RESTful web APIs in Go using the Gin framework
- Learn practical topics such as ORM layers, secure communications, and Stripe's API
- Learn methods to benchmark and test web APIs in Go

Learn Data Structures and Algorithms with Golang
Bhagvan Kommadi

ISBN: 978-1-78961-850-1

- Improve application performance using the most suitable data structure and algorithm
- Explore the wide range of classic algorithms such as recursion and hashing algorithms
- Work with algorithms such as garbage collection for efficient memory management
- Analyze the cost and benefit trade-off to identify algorithms and data structures for problem solving
- Explore techniques for writing pseudocode algorithm and ace whiteboard coding in interviews
- Discover the pitfalls in selecting data structures and algorithms by predicting their speed and efficiency

Leave a review - let other readers know what you think

Please share your thoughts on this book with others by leaving a review on the site that you bought it from. If you purchased the book from Amazon, please leave us an honest review on this book's Amazon page. This is vital so that other potential readers can see and use your unbiased opinion to make purchasing decisions, we can understand what our customers think about our products, and our authors can see your feedback on the title that they have worked with Packt to create. It will only take a few minutes of your time, but is valuable to other potential customers, our authors, and Packt. Thank you!

Index

Made in the USA
San Bernardino, CA
06 March 2020

65376858R00255